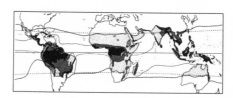

Eco-Geography

Renewal in Science

Eco-Geography

BY

Andreas Suchantke

TRANSLATED BY

Norman Skillen

LINDISFARNE BOOKS

Published by Lindisfarne Books
PO Box 799, Great Barrington, MA 01230
www.lindisfarne.org

Library of Congress Cataloging-in Publication Data

Suchantke, Andreas.
    Eco-geography / Andreas Suchantke.
        p. cm. — (Renewal in science)
    ISBN 0-940262-99-1
    1. Landscape assessment. 2. Landscape ecology. 3. Human geography
    I. Title. II. Series
GF90 .S83 2001
304.2—dc21   00-067172

Printed in the United States of America

10 9 8 7 6 5 4 3 2 1

Contents

The Ecology of Imagination

By Norman Skillen

AT FIRST GLANCE this book might appear to be an example of naturalist travel writing—after all, each of its chapters (except one) contains a portrait of a particular landscape. Though it certainly can be read as a travel book, to approach it in that way would be to miss entirely its essential intentions. These are concerned not only with portraying landscapes but also with a *way* of portraying them, for this book embodies a *way of seeing* that is ultimately derived from the great German man of letters Johann Wolfgang von Goethe. I call him a "man of letters," and that is indeed the way he is best known outside his homeland, but Goethe's life of letters was not only "literary." He also devoted a considerable part of his life to science, and to the development of his own scientific method. As Andreas Suchantke's way with landscapes cannot really be appreciated without some insight into the basic features of Goethe's way with science, I will begin with the latter first. In keeping with the true spirit of the enterprise, I also take a literary starting point.

In or around 1932, in a little shack in upstate New York, the novelist John Cowper Powys—one of the great unrecognized geniuses of this century—sat writing his autobiography. This is an extraordinary book in that it accords as much importance to Powys's momentary sensations as to any "large events" in his life. Thus, for instance, in describing his time at Cambridge he records the following as the most important thing that happened to him there:

> Not far from Trumpington Mill—somewhere in the umbrageous purlieus to the rear of the Fitzwilliam Museum—there stands an

ancient wall; and as I drifted along . . . I observed, growing upon this wall, certain patches of grass and green moss and yellow stone-crop. Something about the look of these small growths, secluded there in a place seldom passed, and more seldom noticed, seized upon me and caught me up into a sort of Seventh Heaven.[1]

He goes on to say that the touch of a pen breaks the spell of this "Seventh Heaven," but he nevertheless defines this intense experience as a "beyond sensation." What Powys here describes is a well-known phenomenon. It is—to use Abraham Maslow's term—a "peak experience," which is very common, *especially* among people in their late teens or early twenties (as indeed Powys was at the time). It figures also in the works of other writers. James Joyce would have called such an experience an "epiphany," and William Wordsworth spoke of "a sense sublime of something far more deeply interfused" and "the pleasure which there is in life itself." However it has been characterized, all these writers seem to concur in seeing this type of experience as spontaneously conveying a sense of the inherent significance of whatever its subject is. It also makes an appearance in the opening chapter of Freud's *Civilization and Its Discontents* as the "oceanic feeling." Freud does not think much of this feeling; indeed he simply discounts it, having no way of taking it seriously. It is, he feels, highly improbable that the world as we know it could communicate inherent significance directly to us in this way. In the Hobbesian/Baconian tradition of science (the main tradition), Freud's has been the standard reaction. We must not be misled by romantic illusions; real reality is harsh, brutal, physical; the only metaphors allowable in connection with it are those of struggle and conflict. Hobbes called it "the war of all against all."

I will not deny that there is something in this view, but why must it necessitate discounting the "oceanic feeling"? What if the oceanic feeling—where "oceanic" may imply depth as well as breadth—is something that has serious epistemological implications for science? What if it could be harnessed, cultivated, applied?

A hundred years before Freud rejected the "oceanic feeling," we already find Goethe speaking of something called *exakte sinnliche*

Phantasie, which in English is usually translated as "exact sensorial imagination." Now, to the Hobbesian positivist—and most scientists even at the end of the twentieth century are still Hobbesians and positivists in one way or another—this must seem like a hopeless, multiple contradiction. How can imagination be exact, and what can it possibly have to do with the senses?

What Goethe meant by this is the practical application of imagination as an instrument of scientific observation. As such it represents a slowing down and a conscious cultivation of the "oceanic feeling." In the oceanic feeling what occurs is a spontaneous expansion of consciousness through which natural phenomena acquire an unaccustomed depth, become charged with meaning, seem to lose their separateness both from each other and their observer, and appear in all their intense relatedness. Powys, for instance, regarded his Trumpington Mill experience as "a prophetic idea of the sort of stories I myself might come to write; stories that should have as their background the indescribable peace and gentleness of the substance we name grass in contact with the substance we name stone." This is what Coleridge called overcoming "the lethargy of custom," when things become redolent of significance which may or may not be capable of articulation. (Where articulation has occurred it has resulted in some of the world's greatest lyric poems.)

In the practice of exact sensorial imagination the same expansion of consciousness, or better still, enhancement of perception, is the aim. To take an example, if we look at the sequence of leaves on an annual or biennial plant as it grows toward blossoming, we will see that each leaf up the stem has a different shape. For the Hobbesian this may be "pretty" but it will not be thought of as anything intrinsically significant (except, perhaps for the purposes of classification). Goethe recommended that we observe these leaf forms until we know them, that we then withdraw and reconstitute our observations in inward contemplation. In this way we can make the formative process between the leaves visible. In imagination we can turn one leaf into the next (something that does not happen as such on the "physical" plant), we can run the whole metamorphic process from root to flower, we also run it backward. Thus the growth process is observable only in imagination, but it is exact and sensorial because based upon concrete observation. The next step in this scientific path is to turn imagination outward again, for the inner practice of exact sensorial imagination strengthens what Goethe called *anschauende Urteilskraft,* perceptual judgment (another contradictory expression). This is Goethe's term for the enhancement of perception that gradually comes with the practice of exact sensorial imagination and results in the ability to perceive unity in combination with multiplicity. We can look at any plant and see the "between" of its different growth forms as one formative process.

This, of course, is a far cry from the mainstream of science where, in spite of a new awareness of the primacy of relationship in ecology and related disciplines, the leading tendency has been toward isolating "basic objects" and viewing phenomena in terms of them.[2] This search is based on the assumption that there are in the universe absolute physical things existing by and for themselves in complete, mindless ontological anonymity, that the properties of these "things" have been and will be the same for all time (the principle of uniformitarianism), and that they are the ultimate basis of reality.[3] The outlook that is a necessary and inevitable consequence of this assumption—mechanistic materialism—has a persuasively pithy rationality about it, but the trouble is that *as a way of seeing* it radically undermines any attempt, be it religious, mythic, poetic, scientific, or otherwise, to apprehend meaning in the universe. No one has stated this in starker terms than Richard Dawkins:

> In a universe of blind physical forces and genetic replication, some people are going to get hurt, other people [or organisms] are going to get lucky, and you won't find any rhyme or reason in it, or any justice. The universe we observe has precisely the properties we should expect if there is, at bottom, no design, no purpose, no evil and no good, nothing but blind, pitiless indifference.[4]

Not many writers have revelled in meaninglessness quite so chillingly as this, but at least this quotation has the virtue of making the position clear.[5] The question, of course, is, What quality of seeing do we bring to "the universe we observe"? What happens to our observation when we begin to make conscious use of its imaginative component?

One thing that this kind of applied imagination discovers is that form is meaning; or, to put it more broadly, nature is experienced as a *language.* Reading this language of nature was what Goethe originally meant in coining the term *morphology.* Hence, for the practitioner of Goethe's method form is not a mysterious, incidental adjunct of invisible molecular processes, but the revelation of meaning, made directly accessible by the power of imagination. If this sounds like a rather round, even brash assertion, it has been substantiated in great detail in a recent book by Henri Bortoft.[6] In chapter 2 he describes Goethe's approach to the phenomenon of color, offering in this connection a critique of materialist empiricism. For instance,

> The error of empiricism rests on the fact that what it takes to be material objects are condensations of meaning. When we see a chair, for example, we are seeing a condensed meaning and not simply a physical body.[7]

Previously he had noted that we do not gain intelligible access to phenomena by purely sensory experience. Purely sensory experience "would be a condition of total multiplicity without any trace of unity"—in fact, in oceanic terms, more like the Ancient Mariner's delirium than his moments of lucidity. Bortoft then goes on to say:

> Since meanings are not objects of sensory perception, seeing a chair is not the sensory experience we [assume] it to be. What empiricism, and common sense, miss through mistaking meaning for matter is the *dimension of mind in cognitive perception.* (pp. 53–54, italics mine)

As the book develops it gradually becomes clear that this "dimension of mind" is imagination itself, and that using it in the recommended way enables us "to enter into the coming into being of the phenomenon," both inwardly and outwardly.

This activity inevitably leads us to a further tenet of the epistemology of applied imagination, namely that form is invariably the expression of a *polarity* of some kind. In other words, it will be a product of opposing but mutually enhancing, even mutually productive tendencies (for example, the language of color, as it is experienced in cognitive perception, arises through interactions of light and dark). The ability to hold within cognitive perception the tension of a polarized relationship is one of the key functions of imagination. It is the ability to appreciate *distinctions* without fragmenting the phenomenon into arbitrary *divisions*.[8] This is what Bortoft calls "comprehensive seeing," the achieved enhancement of perception, which if it happened spontaneously would create an "oceanic feeling," but in its achieved form leads to the perception of "dynamic unity," "the depth of the phenomenon," "multiplicity in unity." So Goethe's approach to science can be construed as using imagination to read the language of polarity in nature.

In the essays in this volume Andreas Suchantke has used this approach to characterize whole landscapes, and it is high time I began paying them due heed. But before I do I would like to clear away any lingering possibilities of misunderstanding what is meant here by "language" and by "imagination." Even if we accept the possibility that imagination working within the process of perception is capable of apprehending form as meaning and therefore nature as language, our understanding of this is still open to interference from "common sense."

One of the chief difficulties of our materialist habits of thinking (everyone has these, no matter what their persuasion) is that they have a tendency toward *reification*—an urge to create physical fixities. In the case of language this urge leads to the assumption that it is an exclusively representational system of arbitrary signs; words are seen as invented chunks of sound that stand for "things." Such nominalism has long reigned supreme, but it has been challenged by a number of modern philosophers, among them Heidegger and, strange as it may seem, Wittgenstein. Besides language as representation Heidegger speaks of *language as disclosure,* which he paraphrases as "saying as showing." The implication of this is that we would not be able to perceive anything around us as meaningful if language had not first supplied the concept. Concepts, therefore, cannot be generalizations from sense impressions—it is just the other way round: sense impressions are only possible because the speaking of the object has caused its meaning to light up in awareness.[9] (Again, without this, sensation would be an unfathomable jumble.) In this way language can be said to have disclosed the phenomenon, and it is this kind of language that is meant in referring to the perception of nature as a language.

As regards imagination there are similar dangers. Materialism, in identifying the mind with the brain, tries to make mental activity, and with it imagination, into a fully localizable, physical phenomenon. What this leaves us with is an epistemologically absurd intelligence (an epiphenomenon of the brain) gazing out as a helpless onlooker upon an ontologically absurd universe (blind physical forces). This perverse state of affairs is a source of daily discomfort, if not distress, to vast numbers of people, and for the life of imagination, as I have been trying to describe it here, it is an unqualified disaster. For imagination cannot be localized (like the mind it is not a phenomenon). Imagination is not a "thing" or a "faculty." It is the mind's propensity to seek unity, to sense the whole that is implied in the part. It is the spontaneous figurative process by which mind, through human perception, makes the world intelligible. As such it is the supreme mediator of meaning to the human soul. This is why Coleridge was at such pains to have it construed as an "active power." It is the activity by which mind and phenomena interact, the tension that lives within the polarity of mind and nature. The practical use of imagination is thus an ecological activity. The quality of this activity determines the quality of humankind's relationship to nature.

This realization, not as a theory, but as a living experience, is the moving thread that runs through all the essays in this volume. They are the fruits of Andreas Suchantke's dedicated efforts on his many journeys to take Goethe's method seriously as a tool of knowledge, to be a living exponent of the ecology of imagination. These journeys were, indeed are, regular interludes in a life that began in Basel (Switzerland) in 1933. There Andreas Suchantke studied biology, after which he became a science teacher at the Rudolf Steiner School in Zurich, a post he held from 1963 to 1982. Since 1980 he has been active in teacher education, chiefly in Europe, but also in South America, New Zealand, and South Africa. Over the years he has undertaken many Goetheanistic-ecological "field trips" to the tropics of Africa and South America, to southern Asia, Israel, and Siberia, as well as constantly renewing his relationship to the landscapes of Switzerland, southern France, and the fields of Scandinavia. His next port of call will be a region of the Himalayas where there is a particularly rich abundance of butterflies. All these travels, entailing a welter of "oceanic feelings" that have been subjected to the discipline of Goetheanist perception, have, of course, produced not only essays, but also a whole series of books (see bibliography) and a large number of papers for scientific journals.

A number of the essays here included represent spin-offs from the books, so that this translation could stand as a general introduction to Andreas Suchantke's work, as well as to his way of seeing. Having already characterized the latter here to some extent, it will be apparent that as a way of addressing the business of scientific investigation it offers a profound challenge, or at least implies considerable modification, of certain strains of biological thinking, while at the same time displaying a distinct kinship with certain other approaches.

On the modification side the Goethean approach generates a view of evolution that considerably circumscribes the role of natural selection in the process as a whole. A direct consequence of the language of form, for instance, is that evolution is not a unidirectional process driven solely by natural selection. In addition to development and adaptive radiation arising out of the selective refining of established lines—essentially a process of evolutionary aging—there is a counter-tendency, which Suchantke, following Julian Huxley, calls *juvenilization*.[10] This forms the subject of one of the essays in this book (the most theoretical one on offer), and he also finds evidence of it in the landscape of New Zealand. It would appear, therefore, that *New*

Zealand itself speaks of evolution as a more complex process than that described by the normal neo-Darwinian narrative.

On the kinship side this book is very much in harmony with new approaches in science that seek to avoid the pitfalls of reductionism by focusing on dynamic relationships both within and between whole organisms—a tendency traceable to the influence of such writers as Gregory Bateson. In Suchantke this again is a direct consequence of his way of seeing, and appears most clearly in his descriptions of various kinds of "threefold organisms" that together form one of this book's major themes. How this view of the organism arises can best be approached by taking an example from an essay that is not included here (although it is available in English).[11] In this essay he arrives by a thoroughly organic sequence of observations at the point of being able to describe the layer of green algae found over the surface of the world's oceans as "the primal leaf." This layer is the interface between the mineral realm of the ocean and the realm of light and air above. This is a strong polarity, whose poles interpenetrate and are unified in the layer of green algae at the interface. The "primal leaf" thus forms the middle component of a unified threesome that spans the whole Earth. This middle component is preserved as a formative principle, which appears as the leaf in the evolution of plants.[12] In the plant the leaf mediates between the mineral realm of the soil and the realm of light and air in exactly the same way as did (and does) the "primal leaf," so that a similar threefold structure is apparent.

In another way a threefold structure appears in the plant in root, leaf, and flower, while in animals it is seen in the polarity between the neurosensory system and the metabolic system, which is balanced by the circulatory system. Then at the ecosystem level these appear respectively as consumers, producers, and decomposers. In the essay on Ngorongoro, Suchantke also sees, within the macroorganism of its animal life, the birds as the neurosensory pole, the ungulates as the metabolic pole, and the predators as the equivalent of the circulatory system. Then at the level of the whole landscape the threefoldness appears again as desert, rainforest, and savanna, where the savanna, as well as being a vegetation type in its own right, can clearly be seen both as desert becoming forest and as forest becoming desert. This has obvious parallels to the "integrative levels" approach to ecology, but with Suchantke it seems to go a step further. Nature comes to be seen not only as a nested hierarchy of integrative levels, but also as a system of—dare I say it?—symbolic correspondences. It is this aspect of nature that bestows upon it the possibility of being spontaneously experienced as intrinsically meaningful. Only on such a basis is the "oceanic feeling" possible. Indeed, in the Ngorongoro essay we have a beautiful evocation of the whole imaginative process from Suchantke's own "oceanic feeling" at the first sight of the crater to his cognitive perception of its animal life as a threefold macroorganism.

Talk of macroorganisms also points to the kinship Suchantke's work has with James Lovelock's Gaia theory. Adherents of the Gaia theory pride themselves on its being a "top-down" approach to planetary ecology, and to a certain extent they are justified in doing so. But there comes a point in any "Gaian" presentation when a reversal will take place and the normal "bottom-up" approach of the neo-Darwinian narrative will reassert itself. How soon this occurs will depend on how "weak" or "strong" the particular Gaian's views are. For "weak" Gaians the systems of the Earth behave *as if* they constituted a single living organism; for "strong" Gaians the Earth *is* a living organism. But weak or strong there will be a cutoff point, and this will usually have to do with the place of *mind* in this giant organism, and more particularly, the place of humankind as the carrier of mind. In this connection mind is likely to be described as an "emergent property" of Gaia. While this represents a significant modification of the Darwinian position, the implication is nevertheless still there that matter came first, and this leaves the "bottom line" of meaninglessness expressed in the passage by Richard Dawkins quoted earlier firmly in position, no matter how much we move the Darwinian goalposts. It seems that no one is prepared to "come out" on this issue and face the ultimate conclusion of the "top-down" approach, which is that if life was primary in the coming into being of this planet and set the conditions for its own development, then mind must be equally primal. This position has been stated by Owen Barfield in the following terms:

Sooner or later a certain truth is brought home to you. . . . I mean the fact that the mind or consciousness is not the function of an organ, though it makes use of organs, the brain among others; that it is not a mysterious something spatially encapsulated within a human or animal skin, but it is the inner side of the world as a whole, just as the individual mind is the inside of one human being. You may prefer in that context to call it "the unconscious." Whatever you call it, you know that, though it makes sense to inquire how and when consciousness developed into what we now experience as such, it makes no sense at all to inquire how and when mind emerged from matter.[13]

The clear implication of this is that evolution is a noetic as well as a biotic process, that it is not consciousness, but the *forms* of consciousness that are emergent, and that the history of this planet is the story of the mutual interactions of these two strands of evolution. This is also a major theme of these essays. It is mostly implicit, but it "comes out" particularly in "The Signature of the Great Rift Valleys," and there I will leave it to speak for itself. In other essays the noetic and biotic strands of evolution appear in the guise of culture and nature. One virtue of this dual view of evolution is that it is a good antidote to the prevalent habit of viewing humankind as an anomalous, freak offshoot of the biological world and its culture as nothing more than a destructive incursion upon an otherwise perfect world of nature. It is also proof against the pervasive notion that nature would be better off without human culture, or that Gaia might one day (soon) eliminate us. This is a philosophy of despair. If mind and nature are the inside and the outside of the One World, then neither of them can eliminate the other in any ultimate sense (although this does not preclude their relationship becoming extremely difficult). Culture, being the conscious pole of this polarity, must learn to attune itself to nature. The possibility of partnership of this kind is another great theme of these essays.[14] It appears especially in "What Have We to Do with the Rainforest?" and "Humankind and Nature."

If Andreas Suchantke is right and the process of juvenilization has created in us the potential to enter into such a partnership with the natural world, this lays a great responsibility on us. It means we have the freedom to develop the sensibilities that meet the needs of the planet—or not. In this we see the full significance of the ecology of imagination. If there is ever to be a viable threefold organism encompassing the polarity of culture and nature, then *imagination* will have the role of mediator. Andreas Suchantke, as these essays bear witness, is someone we might emulate on this score. His life has not been dedicated to carving himself a good academic reputation (which he no doubt could easily have had), but to the loving application of exact sensorial imagination to an incredible range of phenomena, and to capturing the insights gained thereby in words and images. In this he has been as much a poet and artist as a scientist, as his spellbinding descriptions of plants, animals, and landscapes and his exquisite line drawings make abundantly clear.

NOTES

1 John Cowper Powys, *Autobiography,* Macdonald, 1967, p. 199.

2 In the "decade of the brain" that is just drawing to a close, this tendency has been very pronounced.

3 This assumption is very tenaciously persistent, in spite of the fact that in this century physics itself appears to be calling it into question.

4 Richard Dawkins, *River out of Eden,* p. 133.

5 I sometimes have the suspicion that Richard Dawkins is really a member of some esoteric sect trying to put people off materialism—he expresses its implications in such unnervingly drastic terms!

6 *The Wholeness of Nature: Goethe's Way toward a Science of Conscious Participation in Nature,* Lindisfarne Press, 1997.

7 The fact that a chair is an artifact and thus not strictly comparable to animals or plants, which have no immediately identifiable purpose in the way that a chair has, should not distract us from the point that is being made about the *perception of meaning.*

8 The principle of *distinction without division* was first elaborated by S. T. Coleridge in "The Friend."

9 Helen Keller's experience of coming to know the substance water through the word *water* is a famous instance of this. Her account of it can be found in her autobiography.

10 Julian Huxley, *Evolution: The Modern Synthesis,* London, 1942.

11 Andreas Suchantke, "The Leaf: The True Proteus," in *The Metamorphosis of Plants,* Novalis Press, Cape Town, 1995.

12 This is seen as independent of any phylogenetic relationships involved.

13 Owen Barfield, "The Evolution Complex," *Towards,* Spring 1982. Owen Barfield devoted a considerable part of his life to elucidating this question.

14 Andreas Suchantke has written a major book on this subject, so far available only in German; it is called *Partnerschaft mit der Natur,* published by Urachhaus, 1993.

Primeval Past as Living Present

IN THE TIMELESSNESS and the endless fastnesses of Africa some remnants of paradise still persist to this day, or so it seems. Going in search of them is a bit like living out a fairy tale in which dangerous paths must be followed over desolate wastes and the hero must dive into a deep well before emerging into the bright magnificence of fairyland.

Thus, if we wish to reach the enchanted land of the giant crater of Ngorongoro, we must also pass through a series of "trials." Nowadays they are largely only symbolic, but they must have been real enough for the cumbersome expeditions with their long straggling columns of bearers that first traversed this land. Once the sultry forests of the Indian Ocean coast have been left behind, we must thread a way through the endless gray monotony of the thorny scrublands (long weeks of marching for those earlier expeditions) before we pass through the fertile oases and villages of the farming communities at the foot of Kilimanjaro and Meru; then the path leads on across the open Masai Steppe, where earlier travelers were likely to have suffered many a bloody skirmish, until at last we find ourselves standing before a great wall, dark and forbidding.

Without warning the sheer sides of the East African Rift Valley tower up out of the plain like some great rampart piled up by the hand of a giant. Once this wall has been negotiated a further tract of thorny scrub opens out, giving a view of further mountains and the second, higher fault-level of the Rift Valley, graced by forest and cloud-topped hills that seem deliberately placed. Then the path plunges into the swirling gloom of the cloud forest. The enveloping air is moist and cold. Out of the gray swathes of mist loom the ghostly silhouettes of giant figures that look as if they had been frozen in some strange spasm of mirth—the skeletons of ancient trees whose own foliage has long since been replaced by dripping carpets of moss and fern. The climb up through the smothered forest, where no birdcall or other sound is heard, goes relentlessly on. Then a keen wind is felt, which tears the gray veil and blows it away in tatters.

Fig. I. Trees of the cloud forest on the eastern ridge of Ngorongoro smothered in damp moss. Here thick mists form both night and morning. The bushbuck in the undergrowth is scarcely discernible.

Suddenly we are standing seven thousand feet up on the edge of a crater, gazing down into another world, bathed in light and stretching into the far distance. Far below lies the golden-brown floor of the crater of Ngorongoro, framing the mirror of a lake that seems to be not water, but liquid silver. Far behind it spreads the opposite wall of the crater, which in the midday haze looks like a bluish, transparent cloudbank. Not a sharp contour is to be seen, only layer upon light-soaked layer of infinitely delicate color and radiant sunshine in all its glory. As if a piece of heaven had come down to earth.

Fig. 2. The view into the animal paradise of Ngorongoro from the southeastern edge of the crater.

But finding a way down is not so easy. The steep walls of the crater on our side—the south—are covered with impenetrable virgin forest, full of lianas, giant nettles, and army ants, while at other, drier places the grassy slopes are so steep that it would be impossible to find a foothold. Only at a very few spots have steep zigzag tracks been cut into the wall, which can with care be negotiated by a jeep. To get to one of them, however, means walking for miles along the crater's edge, which involves ducking again from time to time into the world of mist and moss and cold. Eventually we reach the place where there is a way down—almost directly opposite our first viewing point. Our surroundings have changed; not only have they become outwardly accessible, but from our new vantage point the light sets all contours in sharp relief. Deep shadows lie in the gullies. What before appeared as purely a play of light and color has now taken on spatial depth.

The scene is breathtaking in its magnificence, its impact increased precisely because it is bounded in the distance. If it were simply an undefined plain like the neighboring Serengeti—which can be viewed from the same spot simply by turning our gaze westward—it would not move us so intensely. And we are not only moved—we are held thoroughly spellbound. To do justice to this landscape, which knows nothing petty or narrow, requires opening oneself to the full: letting oneself be flooded by this power and grandeur is a pleasure without parallel.

Bands of green wind here and there across the plain of the crater floor—a gallery forest, which hugs the stream courses. Where several of these arteries join a small forest spreads out like a carpet, while other areas could well be reed-covered swamplands. In stark contrast to this is the radiant ultramarine of the lake, which in some places is hemmed with reddish pink. Lotus blossoms or flamingos? From the distance it is impossible to say.

Slowly and carefully we drive down the track. At a wet place a cloud of violet butterflies swirls into the air, only to settle again on the same spot. Strange; now they suddenly look quite different, with the colors and typical markings of *Precis octavia*, a relative of the tortoise-shell. The wings have a bright blue shimmer strewn with delicate black stripes, across which runs a band of bright red. If the butterflies are flushed, the air is immediately filled with flashes of violet. Of course! In the fast wing-beats the colors mix!

The further we come down toward the floor of the crater, the more details crystallize out of the uniform brown of the plain. Black points emerge, in dispersed groups or in long chains—herds of wildebeest. Only with the help of strong binoculars and keen observation does it become clear that zebras are also present. Their contours dissolve in the shimmering glare of vapor, dust, and sunlight, black and white merging in a magic flicker that makes the animals literally invisible. All at once zebra markings become comprehensible. They cannot be understood in the zoo, in the wrong surroundings and light conditions. Zebras are animals of wide, open landscapes and long vistas;

their markings were born of the hard light of the plain. That this does not apply only to daylight is confirmed by the observations of the renowned zoologist and Africa expert Colonel Meinertzhagen, who attests to the complete invisibility of zebras in moonlight and starlight, even when they pass by very close.

Thus Meinertzhagen sees the zebra's stripes as a camouflage to protect it from the eyes of its archenemy, the lion, which hunts mostly at night. Simply to establish that this is the case does not, however, go any way toward explaining how such striking markings could come about in the first place. The lion is not the cause of the zebra's stripes (in that, for instance, it gradually ate up all the unmarked zebras so that only striped ones could breed); if that were so, then all species that are prey to the lion, for example wildebeest, would also of necessity be black-and-white striped. No, the protective effect is an incidental, secondary effect.

Fig. 3. Wildebeest and zebras are happy to form herds together.

The attempt is often made to explain a given phenomenon on the basis of discrete factors, but it is also possible to approach it from the point of view of the overall totality of which it is a part. What follows is an attempt at the latter. How do things stand, for instance, with respect to coloring and markings among the other African members of the horse family? Does each form perhaps have its own range? After all, horses occur (or occurred) in every open landscape of Africa except in the west.

It is rather significant, then, to observe that the form with the boldest and broadest markings—Grant's zebra of East Africa—has the most central range, the region around the equator. The dark lines are especially wide and really black, so that they contrast very sharply with the pure white background. They also run right down to the underbelly. With Böhm's zebra, the strain found immediately to the south, whose range runs from southern Kenya through Tanzania, the dark stripes are narrower, and between them there is the pale hint of supplementary markings, the so-called shadow stripes. Further south the tendency to multiply stripes increases with Selous' zebra. Its stripes are narrower, more numerous, and closer together than in the more northerly forms, and therefore the contrast effect they give is weaker—similar, indeed, to that of Grevy's zebra, which lives north of the equator in Northeast Africa and also has very close, narrow stripes. Still further south, from Benguela and Damaraland to the Transvaal and Zululand, lives Chapman's zebra. Here the background color is no longer white, but pallid beige, while the dark stripes are noticeably lighter and the intervening shadow stripes clear and bold. The belly still carries stripes, but the inside of the legs is unmarked. Further south again was Burchell's zebra (*Equus burchelli burchelli*), which used to range from Botswana to the Orange River until it was finally driven to extinction in 1910 by the Boers. It was a "tiger horse" if ever there was one: its basic color was a strong red-brown, the dark lines were blurred, and the shadow stripes were still darker and bolder than those of Chapman's zebra, while all markings had vanished from legs and belly. Lastly, there was the quagga, a denizen of the Cape until its final extermination in 1883. It sported stripes only on the forequarters and

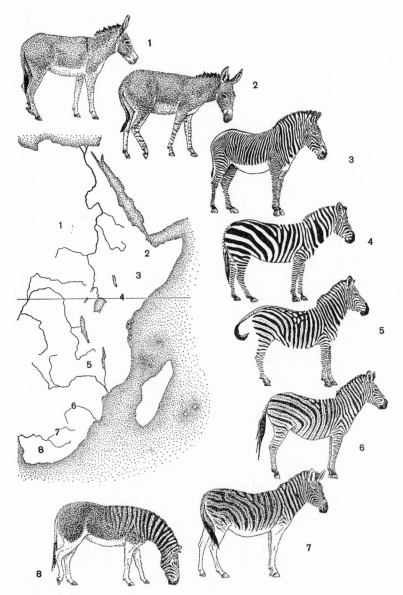

Fig. 4. Distribution of wild donkeys and zebras in relation to coat markings (the mountain zebra, *Equus zebr,* of southern and southwestern Africa is not represented): I. Nubian wild donkey, *Equus asinus africanus.* 2.Somalian wild donkey, *Equus asinus somalicus.* 3. Grevy's zebra, *Equus grevyi.* 4. Grant's zebra, *Equus burchelli granti.* Next in the series—Böhm's zebra, which should come between 4 and 5—has been left out and can be seen among the wildebeest in the previous picture. 5. Selous' zebra, *Equus burchelli selousi.* 6. Chapman's zebra, *Equus burchelli chapmani;* its range extends much further west than is shown on the map. 7. The exterminated reddish "tiger horse," *Equus burchelli burchelli.* 8. the quagga, *Equus burchelli quagga,* also driven to extinction.

neck and had unmarked sandy gray hindquarters—a zebra that turned into a donkey at the back.

Things are very similar north of the equator. Here the continuation takes the form of the African wild ass (*Equus asinus*), which has two strains. The more southerly is recognizable by its black-and-white striped legs and has a range partly overlapping that of Grevy's zebra. The more northerly form finally relinquishes these last remnants of zebra markings—it is sand-colored all over.

Both to north and south, therefore, we see an increasing loss of contrast in the markings of African horses. The initial strong separation of light and dark gradually blurs in both directions, the colors eventually merging. Is there not a parallel here in different qualities of light? Nowhere is the contrast between light and shade so great as at the equator. Here the leaves on a tree can reflect the light as intensely as a metallic mirror, so dazzling that the eyes are blinded by it. Within the branchwork of the tree, on the other hand, deep shadow reigns, so dark and inscrutable that one might be looking into the mouth of a cave. Merciless brightness under the perpendicular sun, deepest shadow wherever it cannot penetrate. The slanting sun, on the other hand, gives the long shadows blurred edges, which merge into the softer light; the slant of course increases the further we go north or south of the equator. At the same time the twilight lengthens—another important moment, when light and dark merge. In the tropics it is so short as to scarcely occur at all, so that every morning and evening there is a dramatic confrontation between light and shade, day and night, just as there is between the black and white stripes on the zebra's coat.

After this short excursion into the geography of the zebra, let us return to the crater of Ngorongoro, for we have not yet properly set foot in it. At one point our attention is captured by a large group of grazing animals of very impressive appearance. For wild animals they are standing too close together and are too variable in color. Then something flashes in the sun guiding our gaze to the spear of a slender Masai tribesman—it is a herd of cattle, and he is the herdsman. The herd makes its way over the plain, getting closer to a group of wildebeest in its path. They do move aside a little, but shortly afterward they are seen mingled among the cattle at the edge of the herd, all grazing peacefully. Cropping the grass close by with similar unconcern is a small herd of Thomson's gazelles. They could almost be goats put out to graze with the cattle—especially the young ones with their leaps and frolics.

Here the primeval past is still present. Thousands of years ago the then-green Sahara might well have presented the same picture, as would many places in Europe during the Stone Age or even later. Human beings were still completely bound up with their natural surroundings, a living part of them, and had not yet elevated themselves into the role of controller, transformer—and destroyer.

Human being and animal scarcely disturb each other. Only at times of great drought is there tough competition around the few water holes. Otherwise no danger threatens the gazelles, zebras, and wildebeest, for the Masai do not hunt—the lion alone is regarded as a worthy opponent for the young warrior, and it is challenged with nothing but a spear.

It must be admitted, however, that even here things have long since departed from their natural state, and a range of delicate measures

Fig. 5 A small flock of Thomson's gazelles, females and kids of various ages. The males have much grander horns.

must constantly be applied to keep the situation in balance. Here as elsewhere this has come about through certain "blessings" of Western culture. In general the Masai do not think much of it, but they make two exceptions and both have the effect of increasing the size of their herds. With gracious condescension, and for horrendous sums of "shillingi" (which they spend on acquiring more cattle), they are always prepared to allow themselves to be photographed; and they object to their herds being tended by vets. As a consequence the National Park Protection Service must constantly resort to new tricks to ensure that not too many head of cattle invade the paradisial grazing grounds of Ngorongoro. But even when there are as many as 1,300 of them, what is that compared to 14,000 wildebeest, 5,000 zebras, and 5,400 Thomson's and Grant's gazelles; the scale of the relationships involved is still more striking when we consider that these figures apply to the rainy seasons, when pasture is plentiful. During the drought, numbers of both game and cattle are much smaller, as the herds wander further afield in search of food.

In view of such large numbers it is no wonder that there are about forty lions in the crater and that there can be few other places in Africa with such a dense population of hyenas. In Ngorongoro, where it is not hunted, the spotted hyena, or *Fisi*, has become diurnal. Everywhere these dirty-brown creatures can be seen lying singly or in little troops before their holes, or squabbling with jackals over a carcass, often only a few meters from herds of grazing or ruminating ungulates. Unforgettable was the sight of a wildebeest bull standing motionless, staunchly guarding his territory, while only a few steps away a hyena lay lounging in the sun, eyeing him quizzically. This peaceful scene between two "deadly enemies" was startling and hard to believe. Contrary to their reputation as cowardly carrion eaters, the hyenas of Ngorongoro are wild and aggressive predators. According to Kruuk's observations they join together in packs and, especially at night, chase zebras and wildebeest, pursuing their victims mercilessly for miles and miles until they give up exhausted and are dispatched with ruthless efficiency. The real opportunists, however, are the lions, in many cases at any rate. They have turned the tables on the hyenas—but it cannot really be called sponging when a lion wades into a howling, ravenous pack and takes the kill for itself. A lioness that had had its lower jaw smashed by a kick from a zebra and was no longer able to hunt, fed itself and its young for a long time on hyena booty, from which it would drive the rightful owners with powerful blows of its paws. Essentially regal behavior, this—the lord has his vassals do the work and enjoys the fruits himself.

Fig. 6. The unappealing physiognomy of the spotted hyena.

Nevertheless, in this landscape there is not much of a sense of deadly enmity or of a struggle for existence. With unhurried ease the parties of wildebeest amble along their trails or lie in the grass. Among them are the bright streaks of bleached bones and weathered, chalk-white skulls; on the bank of the lake lies a fresh carcass from the previous night. And out of the dry grass peep the round heads of young Thomson's gazelles, which could not be more than a few days old. The wildebeest herds are full of half-grown calves, and in a small band of eland at least one cow is heavily pregnant, its belly swollen to ungainly proportion.

A constant succession of birth and death? Even these concepts do not seem to fit. They transpose too much of the human mode of experience onto this world: a world where there is indeed fear, but no despair, pain, but no suffering. Being born and being killed are simply normal parts of ongoing life processes, almost as anonymous as the unfolding and wilting of leaves and shoots on a plant. When a wildebeest is killed, the herd will of course rush off in reckless flight; but before long their pace will slacken, and very soon they will stop and begin peacefully grazing again. Fright, fear, and the missing herd member are not only forgotten, but completely blotted out.

Birth and death of individual animals have really no greater significance than the formation of new cells and the disposal of old ones in the tissues of an organism. It is the organism that persists, taking in substances, assimilating them, and after a time disposing of them again. Here it is exactly the same: the wildebeest is not the single animal, but the totality of all the herds, both large and small. The individual animals do not live separate lives parallel to each other, but are joined together in troops, in bands, sometimes widely distributed, sometimes in tightly packed masses. In long chains, one behind the other, often so close they are touching, they plod along their narrow, well-worn trails that have been used for generations. Some seem to be isolated from their fellows—the territorial bulls, spread out at intervals over the plain and keeping to their posts for days, sometimes weeks on end. But even they are parts of a greater order. It is as if each one has spread out its body in a "diluted form" over the area around its chosen spot, for it reacts to invasion by another bull as if it had been physically wounded, to any infringement of the boundary as if it had been kicked. All in all, then, what we see is anything but an amorphous chaotic mass, even if the herds display no directly visible order.

Fig. 7. Eland, a bull with two cows.

Rather it is a structured whole—the species as an ongoing process—into which the parts, the individual animals, are harmoniously integrated through appropriate mutual relations. It is like an organism that is structured in terms of organs connected by circulatory systems.

The predators also form a part of this organism. They fill the role of the regulatory system, without which the great herds of wild grazing animals would not be able to exist. Strange as that may sound, it is nonetheless true. What would happen if lions, leopards, and hyenas no longer took their tithe? There would be overpopulation, then overgrazing, hunger, and epidemics. In the parts of Africa where leopards have become rare because of trade in their much desired fur, baboon numbers have increased tremendously. But we do not need to look to Africa for examples of the importance of predators. In central Europe hunters have to fill the role of the large (exterminated) predators. If they fail to do this, as during and immediately after the last war, the populations of deer and wild boar get so out of hand that the viability of any kind of rural or forest economy is put in question. Under such conditions, mountain goats in the Austrian Alps multiply so explosively that eventually there is an epidemic of scabies which then all but wipes out the whole stock. An organism is no longer viable if one of its organs begins growing inordinately like a cancer. It ends in collapse.

Predators and herbivores belong together so closely that it is actually impossible to consider one without the other. They are mutually dependent. Where there are large herds of ungulates, lions are also numerous; where there are none, lions are absent. Numerically, however, the predators are always in a tiny minority compared to the herbivores. According to figures and estimates, there are 1,650 lions in the Serengeti National Park (an area about a third the size of Switzerland), whereas there are more than one and a half million herbivores. Taking all predatory animals together—not just the large species, but also mongooses, genets, jackals, and polecats—they make up, according to their "biomass" or weight, 0.6% of the total. How can so few animals have such a large effect? Viewed purely quantitatively, the phenomenon cannot be fully understood.

Balance, of course, does not have only a quantitative aspect; indeed quantity, in this connection, is clearly secondary, a consequence of the qualitative process whereby predators single out the sick and weak for special attention. Numerically these do not matter, and only begin to have a significant effect in this regard if they manage to pass on their illnesses or detrimental features genetically.

Our picture of a macroorganism, a composite unity within which individuals and species are subordinated, attains sharper contours by virtue of this relationship between herbivores and their predators. In this living system the ungulates fulfill the role of the metabolic organs. Ruminants, and also hippopotamuses, are highly specialized in transforming the cellulose of the parched plain grasses into vast stores of tender muscle flesh. The predators cannot do this; their digestion is not powerful enough to destroy dead plant substance and reconstitute it into that of their own bodies. They are dependent on the preparatory work done by the herbivores. The relationship between them is thus mutual as well as predatory.

Predators also do not store the ingested and metabolized substance in their bodies the way herbivores do, but transform it directly into energy and pure activity. After it has gorged itself on a kill a lion or leopard might look fat and ungainly for a while, its belly bloated like a tub. Before long, however, its flanks will have subsided and its form will once again be slender, vibrant, sinewy, muscular. Within the macroorganism the predators are the purest candidates for the role of the circulatory and respiratory system, the level of organization that exerts such a strong regulatory influence upon the metabolic processes —from which it draws its basic needs and which it imbues with the rhythms of breathing and blood circulation. An important feature of this rhythmic system of organs is its mediatory function. It forms the bridge between the two main poles of the organism, the metabolic and the neurosensory. Again and again in Ngorongoro it was brought home to us that this process of mediation is expressed in the behavior of the predators. They are able to swing momentarily between the very sharpest sensory acuity and an absolute, drowsily luxuriant lassitude, an enjoyment of their own heaviness. This appeared particularly in a

Fig. 8. With a threatening look the lioness puts us in our place, then rolls on her back and continues her luxuriant slumbers.

pair of lions that lay sleeping in the grass. The artistic poses in which they reclined bore witness to the care lions take to adopt the most comfortable of all possible positions. When we crept to within a few meters of the sleeping pair, the lioness suddenly raised her head, scratched, and gave us a look of such sharp alertness, of such, yes, regal austerity that we were stopped in our tracks. In that look lay such strength of will—there is no other way of describing it—that we "obeyed" on the spot. Thereupon she rolled happily onto her back, settled all four paws comfortably in the air, and in a moment was fast asleep again. With an alacrity that made the transition imperceptible the animal had gone from sleep to full alertness, and then, just as suddenly, back to somnolent unconsciousness.

Ungulates know nothing of such extremes. They neither wallow in the sensation of their own weight nor achieve sensory alertness to anywhere near what the big cats do. However intense their sensory activity might be, it retains a passive quality. Their sensory activity is pure reflex, totally absorbed in the immediate impressions conveyed to them by eye and ear. The gaze of a giraffe is not penetrating, nor is it fearful, curious, or threatening, but more like a bottomless well. If we remained motionless, even for a quarter of an hour, the large brown eyes under their long silky lashes could remain magically fixed upon us, as if the animal were incapable of doing anything on its own impulse—such as moving to a different spot. It is as if the animal's sensory perception is just as unconscious as its metabolic processes. Its awareness certainly seemed more visceral than cerebral, for if we suddenly abandoned our motionless observer role the legs would be instantly in motion, already driving forward with long purposeful

Fig. 9. Over the top of an acacia she fixes us with an unwavering stare.

strides while the head, half turned aside in indecision and confusion, still held the same position.

Particularly low on the scale of consciousness seem to be those creatures in which metabolic processes are most strongly dominant, above all those with great deposits of metabolic by-products piled upon the housing of their brains—buffaloes and rhinos. Whereas the

Fig. 10. The light, graceful ease of the cattle egrets, and the brooding ponderousness of the old buffalo bull, his horns like a yoke weighing him down.

graceful, nervous gazelles and impalas give the impression that their screwlike horns could almost be antennae, extra sense organs probing the surrounding air, the buffaloes' heads are truly "boxed in." The broad base of the horns appears to press down upon the skull, and the horns themselves do not point outward into their surroundings, but curve downward and turn back upon themselves in a protective gesture that is the exact opposite of that made by the horns of gazelles and antelopes, and echoes the ponderousness of the whole animal.

Observing a herd of buffalo is like watching a brown flood of fertile soil flowing over the terrain. The animals plod along peacefully cropping the dry grass. They devote themselves to the business of feeding, digestion, and excretion with incredible thoroughness, right down to their deep lowing—the sounds they make seem to come not from the throat, but from somewhere beyond the diaphragm. A continual stream of fodder pours into these massive bodies, where it is infused with life in some mysterious way and then flows out again.

Should we come too close to them, however, their heads shoot up, then all movement freezes in a moment, the small eyes staring across at us in all their bewildered dullness. In a gesture that is part repulsion, part helplessness, some of the bulls draw back their upper lip and bare their teeth—not as a threat, but to draw in more air so that they can get the scent better. A single careless movement on our part now produces panic—on a sudden reflex these ponderous colossi begin dashing about every which way and bumping into each other; then they turn into a dense stampeding mass like an avalanche rolling down a mountainside.

Rhinoceroses are so awkward that they even arouse our pity. They have a habit of standing solitary and motionless on the open plain, like their own memorials. Grazing gazelles and zebras amble past, but the rhino does not budge. It is not there for the sake of food, for it is a browser and will retire into the bush at evening. Snorting under the strain, it slowly and awkwardly lowers itself into its hollowed-out bed of sand, first the rump and then the forelegs; then it keels over on its side and lays its heavy head on the ground. In this position it looks from the distance like a block of polished stone. But if it has to it can

Fig. 11. The black rhino.

be on its feet again with extraordinary agility, turning a dim, bleary eye upon its surroundings. Then it stamps, sending up a long trail of dust; then it drops some dung and stamps again, this time with one of its hind-hooves, as if it were trying to bury the droppings. With a quick movement the head is lowered and raised again; it trots a few steps in one direction, then another. If it sees nothing else to worry about, then the whole thing was a false alarm and the rhino sinks again into its favorite state—absolute impassivity. Should the danger come nearer, however, perhaps in the dark, rumbling, stinking form of a motor vehicle, first of all the tail shoots up and curls forward with an amusing flourish (as if this animal, which seemed on the point of turning to stone, had a sense of humor tucked away in one of the folds of its soul). And then, like a rubber ball, this living mountain, which moments before seemed so awkward, rolls across the plain at a speed that all but outpaces our vehicle. With agile ease it gallops along, seeming scarcely to touch the ground, all the while weaving first one way and then another in a zigzag course. All rhinos flee in this way, for their field of vision does not extend behind them and apparently they are incapable of turning their heads, so they have to run in diagonal bursts in order to at least see the danger out of the corner of their eyes. How much of an unconscious reflex this is was shown by a rhinoceros we followed for a long time: although we were not directly behind it but slightly to the left, it still maintained its zigzag course in the "prescribed" way, even though each right turn rendered us invisible.

There can surely be few large mammals with as poor vision as the rhino. Schenkel, a zoologist from Basel who spent a long time studying the behavior of rhinoceroses, reports that they can perceive a human being only within a distance of sixty meters, and then only if he or she is moving. And here once again we encounter an example of mutual enhancement—this time between the rhino and an animal that supplies the missing sensory acuity and is rewarded for this service with rich pickings. The rhino's partner in this relationship is a species of starling, the oxpecker, which lives by removing ticks and tsetse flies from the hides of large herbivores. These gray birds with their bright red or yellow beaks are reminiscent of woodpeckers as they climb around on their lumbering hosts, perching on their backs and heads, keeping a watchful eye on the horizon for any untoward movement and commenting upon it with piercing croaks of alarm. The antelopes, zebras, and especially giraffes, with their more powerful senses, are far less dependent upon the oxpecker than the rhinoceros, for which it literally acts as a substitute sense of sight. Schenkel describes a surprise encounter he had with a sleeping rhino cow. When he had got within thirty meters of the recumbent animal its oxpeckers raised the alarm, whereupon the cow shot to its feet and stormed blindly off—straight toward him. Only when it was a few meters off did his attempts to get out of the way alert it to his presence, whereupon it turned away in fright.

There is a parallel to this association between mammal and bird that is so astonishing it might have come straight out of one of the many African animal fables. As in the former case, a true community of interests brings together the ratel, a honey-eating badger-like crea-

Fig. 12. A female gerenuk. This delicate creature—all ears, eyes and nose—seems to be pure sense organ.

ture, and the honey guide, an inconspicuous brown bird that is a distant relative of the woodpecker. The bird is the one who pinpoints the plunder—a bees' nest somewhere in a hole in the ground or in a tree stump. To get at the desired object, however, it will usually need help, so it flies around in the forest, calling excitedly, until it comes across a ratel. It then flies ahead calling, waits till the "badger" has caught up, flies a bit further calling still, and so on, until it has led its plump associate to the spot. The ratel, with powerful claws and forelimbs as strong as a bear, now clears away all hindrances and noisily eats its fill of honey and grubs. It leaves the combs, which are just what the bird has been waiting for—it is among the few animals capable (with the help of microorganisms in its gut) of digesting wax.

In spite of many differences, the basic features of these two mutually beneficial associations are the same. In both cases the bird takes on the role of sense organ, the mammal that of food supplier. And even though the ratel is not a grazer, it nevertheless is one of the predators in which metabolism is most dominant. As Wolfgang Schad has convincingly demonstrated in his mammal classification,[1] the badgers, these portly omnivores, are the most extreme "metabolists" among the predators—which is readily apparent if they are compared with their nearest relatives, the martens and weasels.

Apart from this exception, however, it is chiefly the large herbivores that find their natural enhancement through the birds. In them the less sharp senses associated most closely with the realm of metabolism are the most highly developed, namely smell and taste. The birds, for their part weak in the sense of smell, have concentrated to an unimaginable degree upon the development of the eye. In other words, they have particularly intensified the sense that is weakest of all in buffalo, elephant, and rhinoceros. And even where the external benefit that bird and mammal have from each other is only minimal or one-way, the fact of mutual enhancement is still undeniable. The graceful, white cattle egrets and quick-moving oxpeckers belong just

---

1. This is based upon the same physiological framework we have been following here, according to which the organism has a clearly identifiable threefold structure consisting of a neurosensory, a rhythmic, and a metabolic system. See Schad 1977.

as much to the dark-brown herds of buffalo as the neurosensory belongs to the digestive system. They are opposites, joined by mutual dependency. The dull, ponderous repose of the grazers has its natural polar opposite in the nervous restlessness of the birds. No sight is more perfect than a cloud of white egrets slowly circling lower and lower over the dark mass of a buffalo herd, until finally coming to rest on the heads and backs of buffaloes. Or an elephant sauntering in dreamy serenity through the savanna while the white birds flit here and there catching insects between its legs or rise in a white cloud, spiraling playfully around the giant.

This picture provides the natural complement to the one we have already attained of the relationship between predators and ungulates. Together all of these animals form one great living organism, in which the birds are the sensory pole, the ungulates and elephants the metabolic pole, and the predators the balancing, regulating rhythmic system. One component of this large-scale organism does, however, take precedence over the others—Africa is without doubt the continent of the ungulates. Nowhere else in the world are such gigantic herds or an even remotely comparable diversity of species found. North America may have had its bison and caribou herds, but it has nothing past or present to set against the vast diversity of African gazelles and antelopes (seventy-one species in all), and the same goes for the Asian steppe. In Africa there were—unfortunately we must also use the past tense here—herds of springbok, of quaggas and other zebra species, of white-tailed and brindled gnus that ran into millions. If this were not enough to demonstrate the "metabolic nature" of Africa, it becomes particularly clear when we compare Africa with other southern continents, especially South America. During the Tertiary period this continent indeed had a diversity of mammals similar to that of Africa, whereas today everything is overshadowed by its colorful bird life and above all its abundance of butterflies.

In Ngorongoro and its relatively close vicinity all the features of tropical Africa seem to be concentrated and intensified. This is most strongly reflected in the crater's high proportion of grassland and savanna, which correspond to the endless plains of the Serengeti, and in the dense cloud-forests we struggled through on our way up the crater wall, which are found on the seaward slopes of all East African mountain ranges. Then, in western Kenya, not a day's journey away, are found West-African type lowland rainforests. Isolated pockets of desert and semidesert with their succulent vegetation are also found here and there, especially around nearby Lake Eyasi. Finally, there is Kilimanjaro itself, a lone mountain the size of a whole mountain range, and a giant living exhibit of all the Earth's vegetation zones from the equatorial to the Arctic desert.

Above all, however, the region that stretches along the Rift Valley from the north of South Africa up to Ethiopia is the setting for the appearance of humankind in a bodily form. In the immediate neighborhood of the crater lie the places where the first finds of hominid and prehominid remains were made: the famous Olduvai Gorge and Mangola on Lake Eyasi. Through them this "heart of Africa" attains a significance far beyond its own continent that renders it the umbilicus, the omphalos of the whole world and the whole of human development. It is our common birthplace. Whether we are Eskimos, Bantus, Europeans, Australian Aborigines, or Americans, we all stem from Africa.

Three Landscapes as a Single Organism

IN THE FOLLOWING SKETCH I will attempt to present the continent of Africa as one great *differentiated whole*. This will involve first describing and then comparing the great landscape types. In doing this, our gaze will be directed chiefly toward the biological and ecological aspects of great living systems such as the rainforest and the savanna.

Because the observations made and the concepts employed might seem a trifle unusual, a few introductory remarks will be made. Readers who wish to plunge straight into the landscape descriptions may skip over them.

I. THE ORGANISM CONCEPT APPLIED TO LANDSCAPE AND CONTINENT

Let us begin by bringing into sharper focus what is meant by landscape in an ecological sense. In normal parlance, *landscape* refers simply to whatever panorama offers itself to the gaze of an onlooker standing at some point on the Earth. No matter how sweeping this onlooker's gaze, it is unlikely to take in any kind of complete unit, but only an arbitrary extract determined by the "point of view." In spite of this arbitrariness, however, the landscape will be perceived as consisting of distinct, coherent structures of variable extent, each with its own unity—forests, rivers, lakes, beaches. A closer look will reveal that each of these unities in turn has its own inherent structure, down to the smallest detail, but all these single components or regions in combination merge into the totality that is the river, the forest, the steppe, the moor. Landscape in this sense, then, is a clearly contoured phenomenon, whose fine detail is fitted into larger structures. These higher-level structures imbue a given landscape with its characteristic signature or physiognomy.

The expression of this signature, however, is not confined to the large-scale features of a landscape, but lives also in what might appear to be its smallest and most insignificant details. The rich upholstery of moss and the large, delicately textured leaves of the herbaceous plants and ferns that form the typical scene in a moist, shady forest would never be found in rocky landscapes exposed to direct sunlight, extremes of heat and cold, and long periods of drought. In such places a key contribution to the look of the landscape is made by single spongy, ball-shaped bushes growing in clefts, and scrubby drought-resistant shrubs with small tough leaves.

What we are seeing, then, in the face of a whole landscape is what might be called its *ecological gestalt,* the visible expression of the integrated interplay of all its living and nonliving parts. This in turn is subject to a further complex of forces, namely the climate, which itself is the expression of the still more comprehensive interaction of Earth and Sun.

From landscape as "ecosystem" it is only a short step to landscape as "organism." Indeed, almost all the distinguishing features of an organism apply also at the level of the ecosystem, assuming, of course, that we have a conceptual framework flexible and comprehensive enough to handle such connections. Here natural science leaves us somewhat in the lurch, for it recognizes no principal difference between the living and the nonliving, and consequently has not developed terminologies specific to particular kingdoms of nature. Among the few thinkers who have approached this question rather differently is Rudolf Steiner, who considered the nature of the organism repeatedly from one perspective or another, seeking to clarify its key features. From these deliberations *seven processes* were identified, which together make up the life of an organism. If, therefore, an ecosystem is as much a living system as is an organism, we could expect to find the organism's "septet" of key functions developed into an "oratorio" at the ecosystem's level of organization.

Let us consider them one by one. *Respiration,* which occurs in all living organisms, expresses itself at the ecosystem level in the constant stream of CO_2 emanating from the soil, or in the large-scale metabolic exchange of oxygen. The ecosystem equivalent of *thermal regulation* can readily be discerned in the uptake of solar energy. But what of *nutrition,* the intake of substance from the environment? Ecosystems, after all, *are* the environment, and when they are stable all substances are caught up in cycles of assimilation, excretion, and reassimilation. There is, however, one key exception to this, namely, water, which continually flows in and out of the system—from or to the atmosphere or the ground—and without which none of these other processes is possible. Ecosystems, we could say, therefore, feed on water.

In addition to these three basic processes in which substances enter the organism from outside, there are processes that take place within the organism itself. In *decomposition,* substances are broken down, either so new ones can be absorbed and built up, or so old ones can be excreted. In the ecosystem, the complex interchange of substances along the manifold food chains could be seen as corresponding to this basic metabolic activity. *Conservation* expresses itself in the organism in the way morphology and somatic balance are preserved in the face of a continual throughput of substance. In the ecosystem, on the other hand, it appears as ecological balance, the faculty of self-regulation, which keeps the variation of population numbers within bounds, as, for instance, in the suicidal migrations of lemmings. But can an ecosystem be said to show *growth* as organisms do? In a small number of cases we have evidence of definite life cycles with juvenile and growth phases leading to mature and even senescent stages, for example in moorland ecosystems. In other major landscapes such processes are so slow as to be scarcely observable, except where chance events have treated us to the spectacle of emerging ecosystems, as on newly formed volcanic islands such as Surtsey (Lindroth and Schwabe 1970), or where successive glacier regression has placed the different phases of a living community side by side for our examination, as in the case of the Aletsch Forest (Richard 1968). And what of *reproduction*? The comparison becomes meaningful when we think of this not as mere numerical increase but rather as propagation: in other words, as the continuity of the species and population over and above the individual organism, which is only a tiny episode within the total process of the ecosystem it is embedded in. It would seem, however, that at the ecosystem level there is not much to choose between the three processes of conservation, growth, and reproduction. Here distinctions seem just as artificial as in simple organisms, where the border between conservation and growth (during regeneration) is as fluid as that between growth and vegetative reproduction.

Organisms, however, have still other basic features that also find expression in the structure and functioning of living communities. The first of these is *rhythm*. The rhythms occurring in living processes display an inner and an outer or, if we prefer, a microcosmic and a macrocosmic aspect. Whether endogenous or organ-specific, they are not merely synchronized with cosmic-planetary, particularly Sun-Earth, rhythms, but seem quite simply to be their organic continuation. This comes most clearly to light when an organism is placed in totally artificial surroundings cut off from all cosmic rhythms—life processes become chaotic and disintegrate. The inter- and intraspecific relationships in an ecosystem correspond to the "endogenous" rhythms at the organism level; this can be seen, for instance, in the way relationships between plants and animals in the savanna change with the transition from dry to rainy season or in the way some flowers open at quite specific times of the day to be visited by pollinators such as moths and bats that become active at exactly the same time. In the attunement of its "inner" rhythms to the diurnal and seasonal cycles, the ecosystem is almost as precise as the individual organism.

A further basic feature of living organisms is *functional differentiation*. The course of phylogenetic development displays a gradual tendency for organisms to "separate out" certain basic life processes and form them into distinct organs, while at the same time intensifying their mutual dependence. (Crystals display similar formative tendencies, but they are denizens of the inorganic realm, so their parts, in the absence of impurities, are all completely identical.)

Functional differentiation doesn't occur willy-nilly, or in a separate way for each species, but follows distinct patterns. Perhaps the most important and universal of these is the principle, discovered by Rudolf Steiner, that *physiological function is organized into three systems*. According to this principle the life of a complex organism consists in the activities of functionally opposed organ systems. This polarity is integrated into the organism that gave rise to it in conjunction with a third, mediating, system, which brings the two poles into synergetic interaction, making them complementary rather than antagonistic. Steiner developed this principle of threefold organization using the human organism as an example (e.g., 1917). Here the primary polarity is between the central nervous system (the "animal pole" of embryology) and the metabolic system (the "vegetative pole"). The denaturing effects of the former is opposed by the synthesizing power of the latter, and the circulatory system mediates between them, supplying the nervous system with the products of metabolism while at the same time purging the body of the harmful by-products of neural activity. The full extent of the opposition between the two poles, however, is revealed only when evolutionary development reaches a certain level of complexity. As functional differentiation increases and the central nervous system becomes more highly developed, vegetative capacities decrease, for example asexual reproduction and regeneration.

This concept of threefold organization has already been fruitfully applied to living communities and large ecosystems and as such needs no special justification here. It suffices to mention several studies that have successfully followed up Rudolf Steiner's lead; for instance, comparisons of different types of lakes by Gümbel (1972) and of the various stages in the course of a river by Koehler (1971), and the description by Grossbach and Schad of the phases involved in the formation of moorland (1974). A number of purely ecological studies have come to very similar conclusions. Thienemann, in his well-known study of a lake (1956), gives a picture of an organismic whole extending over three different areas—the shore region, the open water, and the deep water. Each of these three zones is inhabited by a different group of organisms, each of which in turn performs a different function within the whole. Thienemann observes, moreover, "We find such groupings not only in lakes, but also over the whole earth." The shore region is the domain of the green plants, which, through their assimilative capacity, synthesize biomass. They are the producers, upon which the animals of the open water, the consumers, both herbivorous and predatory, live and feed. The consumers transform the organic matter they ingest in a variety of ways and perform, through excretion, the first stages of the process of decomposition. The actual process of decomposition, however, takes place in the depths of the lake where countless microorganisms, the decomposers, "break the complex organic compounds down into their original elements and in this way channel them back into the general chemical cycle. . . . The cycle of elements binds all three ecological groupings into a higher unity that is the lake itself." Thienemann then concludes: "The fact that even in a living community with such a simple threefold structure we encounter phenomena with features that are not comprehensible in themselves but only in the light of other functional components of the community—in other words, the fact that such a community represents a living whole—is an insight of no small significance. For if the essential features of even the simplest community are understandable only insofar as they are seen to reflect the holistic nature of an ecosystem, how much more necessary is such a 'holistic approach' to the intermeshed workings of the communities that make up the whole terrestrial cosmos, if we wish to attain deeper knowledge of it."

It may nevertheless be objected that in spite of all evidence to the contrary there are still such basic differences between organism and ecosystem that no correspondences between them can be taken seriously. Debate on this point has been hard fought since the 1930s, when Clements and Tansley first propounded the concept of the superorganism. Detractors (Remane 1950 and, following him, Tischler 1955 and others) point to the fact that the organism is characterized by functional coordination, the ecosystem by opposition between its parts. We have already seen, however, that opposition is a fundamental feature of the organism, while coordination—in the form of synergy—is also a familiar part of the picture.

A further objection raised is that the community, for example the plant association, is not a distinct, necessary entity but a contingent patchwork of individual species, in contrast to the organism, whose integrated structural elements all developed out of a common rudimentary form. Just how artificial this distinction is, is shown by the abundant examples of intimate association between organisms—for example the complex relationships between blossoms and quite specific pollinators—none of which could ever have come about without a long period of complementary morphological development, in other words, coevolution. In extreme cases of highly developed symbiosis, the coevolving organisms may be so closely bound up with each other that they give the impression of a single organism, as in the lichens. But even if we take as our example the broadly based interdependence of producers, consumers, and decomposers in the general economy of nature, we can have no grounds for supposing that it came about through organisms developing in isolation and only entering into relationships with other organisms once their own specialized development was accomplished. They have all traveled long paths of mutual interaction. A living being with no place in the community of life is just as unintelligible, just as unviable, as an organ cut from its parent organism or a cell in isolation. And if we go back far enough into the beginnings of animal and plant evolution, we eventually arrive at a single, unified rudimentary form, which has the same relation to the highly differentiated life-forms of today with their complex, ecological interconnections as the embryo has to the mature organism.

It must, of course, be admitted that there *are* differences between organisms and communities. The purpose here is not to try to minimize them or to argue them aside, but rather, while acknowledging their differences, to point out the striking similarities in their basic features. The fact is that on different levels and in varying dimensions we encounter the same phenomenon. The analytically inclined will perhaps lay greater emphasis upon the differences, the synthetically inclined upon the similarities. A truly appropriate response to the nature of the organism, however, lies in seeing these two modes of thinking as complementary (Vogel 1972). A synthesis without exact analysis remains vague and superficial, whereas pure analysis delivers only in-depth information on unconnected fragments instead of integrated knowledge.

Whatever the scale of ecological communities, then, whether pond or lake or whole landscape, they can be said to have the same basic features and follow the same underlying principles as organisms. (Not surprisingly, we find that the ecological geographer Schmithüsen calls landscapes "synergetic communities.[1]) Small-scale communities exist within the context of larger ones, ultimately of "landscapes," those eco-geographical entities with which this essay is concerned. It is clearly apparent that in the living world, as in the rest of nature, the same basic structures are found connecting the various levels in a nested hierarchy in which each part is the emblematic complement of the whole. The complementarity of part and whole is reflected at every level: a living creature consists of organs and is itself an organ in a higher organism, the living community, and so on.

In what follows the attempt will be made to show how higher order living systems—ecosystems—are themselves organs of a still greater whole within which they perform certain functions in cooperation with other ecosystems. If Africa is the object of study and a threefold structure once more emerges, this would seem to suggest that even a continent is not just a coincidental spatial aggregate, but a functionally differentiated whole, an organism.

Applying the organism concept to a whole continent, however, needs fuller justification. Here, unlike in an ecosystem, our first impression is not of a self-consistent entity with parts subject to the influence of the greater whole. A continent stretches through various climatic zones and incorporates a whole range of large-scale ecosystems. Would it not be better to view these gigantic systems, each with its own characteristic climate and its own typical populations of animals and plants ("zonobiomes," Walter 1976)—for example, the tropical rainforest with its diurnal climate, the monsoon belt, or the subtropical desert region—as organlike structures of a still greater

---

1. The German word here is *Synergose*, which is reminiscent of the term *Biozönose*, an alternative term in German for a small ecosystem or biome. —Trans.

whole, namely the whole Earth? These, after all, are not confined to single continents, but span the whole planet. True as this is, it does not alter the fact that each continent has its own version of each of these great ecosystems. Each continent makes its contribution to the worldwide system, but at the same time makes it "its own." There is thus a typically African rainforest and a typically African savanna, both standing in sharp contrast to their South American and Asian counterparts. More will be said of this in what follows. Furthermore, these large landscape types vary in the weight of their impact within their respective continents, In South America, for instance, the deserts are nothing like as significant as they are in Africa, where on the other hand moist forests are marginal in comparison to those of the neighboring continent in the west (Fig. 1).

Each continent, then, represents a unique composition of the Earth's major landscape types. This sketch of Africa is intended to show that such compositions are not mere geographical juxtapositions, but that regulated interaction in obedience to the principle of complementary functional differentiation is much more the order of the day.

II. THE LANDSCAPES OF AFRICA

From the "life pole" of the equatorial rainforest to the "death pole" of the desert, Africa encompasses tremendous contrasts of landscape. Between these extremes the savanna forms the intermediary zone. It is so full of unique characteristics and so teeming with life that one need have no hesitation in calling it the heart and crown of the landscapes of Africa. At the edge of the continent there are two further regions with which this essay will not be concerned: in the north, the Mediterranean zone, especially the Maghreb, with its tough-leaved vegetation; and in the south, at the furthest tip of the continent, the comparable but, in its combination of plant species, totally unique landscape of the Cape Province. Both are characteristically un-African, being the two single regions of the continent affected by more recent rock folding. In both areas also, Africa in the form of the large animals has been driven back by humans and largely eliminated. Moreover, in terms of human geography and culture both these marginal regions—North Africa since ancient times, South Africa since more recent—have been more strongly subject to foreign influences than any other part of the continent.

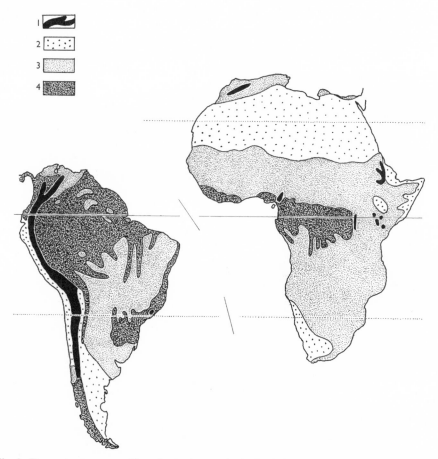

Fig. 1. The contrasting composition of natural vegetation in South America and Africa. Whereas in South America the periodically dry savanna-like landscapes are surrounded by forest areas, the opposite is true in Africa.

1). Alpine vegetation; 2). Desert and semidesert; 3). Savanna and arid plains vegetation with distinct rainy and dry seasons, including contrasting landscapes from grasslands to dry forest. In South America, from south to north: *pampas, chaco, cerrado, caatinga*. In Africa, as in Fig 2. ; 4) Wet forests; equatorial rainforests and, in South America, Atlantic forests (recently these have dwindled to small areas along the coast), Patagonian and subandean forests. (Africa from Keay 1959 and Moreau 1966; South America from Hueck 1966 and Eiten 1974.)

Our sketch must also leave out the African alpine landscapes, together with the marginal habitats of the coast, chiefly the mangrove swamps, which are of little significance for the total picture of the continent. This is something of a pity for the alpine regions also display a threefold structure: the zone of moist forest; then, above the tree line, the "savanna" of grass flora loosely interspersed with giant lobelias and treelike *Senecio*; and finally the "desert" in the scree zone. It is just like a scale model of the whole continent, except with different flora and fauna, and would have been well worth a study of its own (cf. Coe 1967; Hedberg 1951, 1961; Troll 1959).

Deserts, rainforests, and savannas are also found in other parts of the tropics. Nowhere, however, is the savanna so ecologically significant as it is in Africa. In South America the richest diversity of flora and fauna is found in the rainforest. In Africa the situation is different. If the areas of the land surface occupied by the three main African landscape types are compared, the savanna—in its widest sense, ranging from dry forest to semidesert (in the appropriate section the nature and extent of the savanna will be defined more closely)—covers the

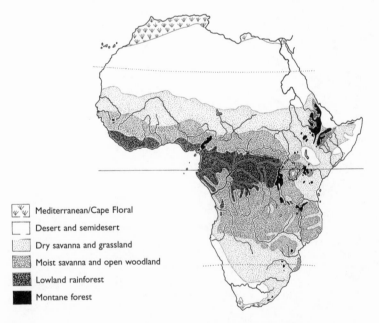

Mediterranean/Cape Floral
Desert and semidesert
Dry savanna and grassland
Moist savanna and open woodland
Lowland rainforest
Montane forest

Fig. 2. Main vegetation zones in Africa (from Keay 1959 and Moreau 1966).

largest area. Second is the desert, while forest, both lowland and upland, comes last. The savanna, however, stands out not only in spatial terms, but also in the abundance and diversity of life-forms it supports. It will therefore be treated at greater length in the description of the bioregions that follows, while forest and desert will be characterized in terms of a few essential features.

The Rainforest

The relatively small area occupied by rainforest in Africa runs in a narrow band, interrupted here and there along its length, along the coast of Guinea and widens out in the east into a somewhat large expanse that stretches over the Congo basin as far as the Central African mountains. Isolated patches of rainforest, diminishing the further east they are, still exist east of the Central African lakes in Uganda (Budongo, Kayonza, Mpanga; cf. Moreau 1966, Reynolds 1966, Williams 1971), in western Kenya (Kakamega Forest), and in narrow bands of semiwild forest remnants along the East African coast.

The moist upland forests bear a fairly close resemblance to the rainforest, but the composition of their flora has a characteristically higher proportion of genera from more temperate northern latitudes (e.g., juniper [Juniperus], larkspur [Delphinium], anemone, lady's mantle [Alchemilla], bramble [Rubus], violet [Viola], cranesbill [Geranium], and countless others; cf. Agnew 1974, Hedberg 1961). Interesting features are displayed by the forests on the lower slopes of isolated mountain outcrops in the coastal monsoon belt of East Africa, especially those of the Usambara Mountains in northwest Tanzania. Their flora betrays a strong connection with West Africa, even though the two regions are separated by the vast arid fastnesses of the savanna. The connection is even more apparent in the fauna and remains detectable in the birds and insects in other pockets of upland forest in East Africa (Carcasson 1964). These and other related phenomena are evidence that, apart from shrinkage during the Ice Ages, the forest covered an extremely large area. Since the last Ice Age and in the two previous interglacial periods it probably stretched unbroken from Ethiopia over the highlands of East Africa right down to Angola and South Africa (Moreau 1952, 1963; van Zinderen, Bakker, et al. after Walter 1973). Its extent can be reconstructed from numerous animal and plant remains and coincides fairly closely with the region in which the ranges of forest and savanna-dwelling mammals (which of course occur in mosaic fashion) currently overlap (see Fig. 3).

Let us now begin our journey through the landscapes of Africa in the rainforest. At ground level all is humid gloom, the forest floor covered in rotting leaves. This leaf debris is every bit as thin as the rhizome-rich layer of humus below it, which, if scratched away, reveals the sterile red soil immediately underneath, together with the dense mat of roots tangled through it. The going in this forest is mostly quite easy. Nowhere do we find the impenetrable "jungle" we expected. This is sooner encountered at the forest edge or in patches of savanna dominated by acacias, where the way will be barred by a green wall of thorns as high as the trees themselves. Pale column-like trunks with thin, smooth bark reach up, without bothering much about side branches, into the heights, where their branches and leafy crowns merge into a sea of green: of many trees only the trunks are visible. Time and again we come upon truly gigantic trees that bar our way

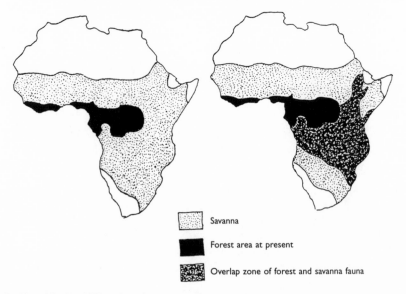

Savanna

Forest area at present

Overlap zone of forest and savanna fauna

Fig. 3. (From Kingdon 1971, redrawn.)

Fig. 4. Kapok tree, *Ceiba petandra*, with powerful buttress-like roots that reach high up the trunk; Congo rain forest (from a photograph by Rahm 1973).

with the splayed-out buttresses of their roots. They seem to be standing on the earth rather than rooted in it. And, indeed, underground the roots hug the surface so closely that here and there they reemerge, coiling over the earth like great snakes.

Gradually the inescapable sense grows upon us that what we are walking through in this world of the forest floor is actually the root zone of these giant plants. Everything we see is rootlike: the hanging ropes of the lianas are scarcely distinguishable from the ubiquitous strangler fig and epiphyte tendrils, which are actually roots. Truly it is as if we had found our way into the underground realm. For a comparable impression in more temperate climes we would need to enter the rank gloom of an overgrown gorge. The earth's surface seems to be up there where the light falls on the spreading tops of the trees and is swallowed by the foliage or reflected by glowing blossoms, only the merest fraction of it penetrating to the ground. In comparable forests in Sumatra, values of between one-thousandth and two-thousandths parts have been recorded (Bünning 1947). The oppressive feeling of being hemmed in that we experience in the rainforest comes not from any lack of space, but from being cut off from the light and enveloped in a vegetative realm of intense, mute, unconscious growth. Subjective as such an impression might be, it points to a real, objective danger to which anyone from more temperate parts who stays for any length of time in the rainforest is prey. The brooding closeness of the atmosphere, which persists even at night, and the eternally repeated daily alternation between dizzying heat and torrential rain dull the mind and smother all activity. A slow, inexorable fall into inertia sets in, leading ultimately to a complete disintegration of personality. Conrad's "Mistah Kurtz" is not the only one to have suffered this fate. Not for nothing are many areas of tropical West Africa called "the grave of Europeans."

Gazing up out of the gloom toward the forest canopy, the eyes are met by myriad tiny windows of light framed by the black scissor-like patterns of the leaves against the blazing sky; the songs and piercing cries of birds echo down, and the drone of beating wings is heard when great hornbills sweep over the treetops. They are just as invisible as the fleeing troupe of meercats, which betray their presence only through the rustling and crackling of branches accompanied by bird-like twittering. Looking down again, we see a large fiery-red blossom lying at our feet, a sign that we are standing under a Nandi flame tree

Fig. 5. The herbaceous layer of the tropical rain forest is found primarily in the crowns of the giant trees that rise above the dense forest canopy. "Hanging gardens" with arboreal soil deposits develop on branches at a height of 40–50 meters. Many of the plants that grow here collect humus and possess specialized organs such as protective leaves that attach themselves to their base. In addition there are carpets of moss and the soil built up by the epiphytes sustains other plants growing on the trunks and lianas. The humus layer on the ground, on the other hand, is extremely thin (5–10cm). The roots of the great trees hardly penetrate further into the underlying nutrient-poor mineral soils. (Compare Fig. 7 in chapter 3, "What Do Rainforests Have to Do with Us?")

(*Spathodea campanulata*). The crown of this tree must be alight with blossom, but of this glory we see nothing.

In many ways these upper regions far beyond the reach of our eyes represent the equivalent of something that in temperate forests we normally see by looking down—the herbaceous layer. In the rainforest many of the smaller plants grow not in the ground but as epiphytes in "hanging gardens," which have accumulated on branches and under other epiphytic growths, together with a full complement of mites, springtails, and even earthworms. Ants, exclusively ground-dwellers in temperate forests, here prefer to build their nests in branches that stick out a little above the forest canopy. Many species of termites follow their example. With the grazing mammals the pattern is the same. Few are found on the ground, for example duiker, alone or in pairs, or the rare Bongo antelope, while large troupes of colobus monkeys swing through the high branches. Herbivorous apes are a characteristic phenomenon of tropical rainforests, and not only in Africa. In South Asia langurs (*Presbytis*) fill this role, in South America howler monkeys, as well as sloths. Particularly striking also is the fact that in both their mode of life and their physiology there are parallels between these inhabitants of the treetop meadows and the ruminants (Bauchop and Martucci 1968; Goffart 1971).

The forest floor where we walk is the region of decomposition—and of germination. The latter occurs less in than on the ground, since the soil is poor in minerals and has an extremely thin humus layer. All the important minerals for plant life are caught up in cycles largely above ground. Any that get into the ground are washed away by the rain. This explains why a tree can sprout only if the seed lands in the rotting remains of a fallen one (Richards 1964, 1973).

To disregard this fact brings catastrophic consequences, all too familiar in connection with attempts to clear the rainforest for agriculture. When the land is cultivated after felling and burning the trees, the initial crop yields are very good, for the soil has been enriched with minerals—a highly deceptive imitation of temperate conditions. The situation changes rapidly, however, when the sun, no longer hindered by the forest canopy, shines on the bare earth, burning up the humus, and the soluble minerals are leached out of the soil. All that is left are insoluble aluminum and iron compounds, which in extreme cases harden into a rocklike crust (McNeil 1964). The ruined land is now only capable of supporting sparse, drought-resistant plants, mainly grasses, and is lost as much to humans as to the forest, which can never recolonize the exhausted soil. In the rainforest there is no chance, as there is in temperate climes, of preserving fertility by careful husbandry or of the forest returning once cultivation stops. To destroy the world above ground *is to destroy everything*. Rainforest cultivation can therefore be justified only if it is integrated into the forest by thinning out the trees instead of displacing them.

In the rainforest, animal life-forms are completely overshadowed by the dominance of the plant world; indeed the higher their level of development the more in retreat they seem to be. The insects, by contrast, are superabundant. Not only are they represented by a vast number of species, but they also occur in their largest forms (for example, the Goliath beetle). Among these are the large forest butterflies. Those in Africa, apart from a few exceptions (e.g., *Papilio antimachus*), cannot match the wingspans of their South American relations, and the colors of their markings are also more muted.

A significant feature of the African rainforest here comes to light. The high level of development and the great diversity of form, especially in the plant and insect worlds, displayed by the tropical forest areas of other continents, above all South America, are not present in the African forest to the same extent. In the animal world this is apparent in the large butterflies and the birds (see what is said on this topic in the section on the savanna); in the plant world it applies both to forms typical of the rainforest such as epiphytes and lianas and to numerous other taxonomic groups, which in comparison to the forests of the Amazon or Southeast Asia are weakly represented. Particularly striking is the paucity of palms. In the whole of Africa there are only about fifty species, whereas both in South America and Australasia there are over eleven hundred. The same contrast with other areas is found in the orchids. They are less eye-catching, both in terms of the variety of species and in their marked preference for small blossoms.

Nevertheless, the African rainforest is still the zone where—even in purely quantitative terms—the plant world in its luxuriant fullness rules over everything else. Closely connected to this is another phenomenon that, in line with observations already made, appears at its most striking in the African rainforest. This is the prevalence of dwarfism among the higher life-forms, the mammals and the forest-dwelling humans. The forest strains of the African buffalo (*Synercus caffer nanus*) and elephant (*Loxodonta africana cyclotis*) are noticeably smaller than their relatives of the savanna (*Synercus c. caffer, Loxodonta a. africana*). Particularly among the ungulates we find many small and dwarf species, from the pygmy hippopotamus (*Choeropsis liberiensis*) and the duiker (*Cephalophus*) down to the house-cat-sized chevrotain (*Hyemoschus aquaticus*) and the tiniest buck of all, the twenty-five-centimeter-high royal antelope (*Neotragus pygmaeus*)—a Lilliputian by any standards. Among the apes there are also distinctly smaller forms, such as the pygmy chimpanzee (*Pan troglodytes paniscus*), which differs from the larger races of chimpanzees in that it keeps exclusively to the rainforest (Dorst and Dandelot 1972). And finally humans appear also in a dwarflike form much more pronounced than in the rainforests of other continents—the Bambuti Pygmies.

It is significant that the dwarfism found in both human beings and higher animals in the rainforest is not simply a reduction in size that preserves normal proportions. It produces juvenile bodily features. Pedomorphic traits are clearly apparent when we compare, say, the pygmy hippopotamus with its larger relation. It is shorter in the snout, and the head is more rounded, "like a baby." The horns of the dwarf forest buffalo stop growing at a stage of development that the buffalo of the savanna goes through as a juvenile; the proportions of the short-necked okapi resemble those of a young giraffe, etc. (Fig. 6). This tendency is particularly pronounced among the Bambuti peoples, above all the Ituri Pygmies, in whom, "in contrast to other peoples, there is a distinct lack of growth during the years of puberty" (Martin and Saller 1959). In their legs, which are especially short in relation to the torso, in their high, protruding forehead, and in their fine, soft body hair, they display childlike features (Schebesta 1938; Schwidetzky 1974).

It is no surprise that the *world of insects* finds ideal living conditions in the rainforest. This is shown by the high degree of morphological convergence between insects and the plant world. Perhaps the clearest example of this is the butterfly, with its broad wings not only open to the light but also permeated by it. Organs that stretch out into

Fig. 6. Closely related forms or races of the same species native to the savanna (left) and the forests (right): chimpanzee *Pan troglodytes* and pygmy chimpanzee *Pan (troglodytes) paniscus* (from Dorst and Dandelot 1972). African buffalo *Syncerus caffer caffer* and the smaller dwarf forest buffalo *Syncerus caffer nanus* (from Fauna, vol. 3). Hippopotamus *Hippopotamus amphibius* and the pygmy hippopotamus *Choeropsis liberiensis* (from Dorst and Dandelot 1972).

the surrounding world and the formation of large surfaces open to the influences of the environment are not features characteristic of animal organization, but belong much more—in the form of leaves—to the plant. And indeed, tropical forest butterflies receive the markings and coloration of their wings from the all-pervasive interplay of color, light, and shadow in their leafy habitat (Suchantke 1974, 1976; Papageorgis 1975).

In the *warm-blooded animals* we are presented with just the opposite picture. Their bodily organization both incorporates and gives outward expression to what lives within them in the way of mental processes. This turning inward of the organism goes hand in hand with a certain degree of emancipation from the environment, but involves too a distinct inhibition of the body's regenerative capacities. As was mentioned in the first section, the richer and more inwardly differentiated mental process becomes in the course of evolution, the more the faculties of regeneration and vegetative reproduction diminish. For organisms at this level of organization, failure to achieve physical maturity could be taken to imply that this process of inward change is correspondingly incomplete. For *humans* this could mean not reaching the stage of full, individual self-consciousness. The child, after all, is distinguished from the adult in that those processes exclusively active in plants dominate—the processes of growth and form.

But let us look into this in a bit more detail. To say that the Pygmies are like children is not necessarily pejorative—this feature is not restricted merely to a premature arresting of growth just before puberty. From the very outset their childlike features are more pronounced than in other peoples; for example, their infants tend to have larger heads (Schebesta 1938). Puberty as a developmental process is a complex interweaving of many layers of the growing child's bodily and psychological makeup. The bodily processes chiefly affect the reproductive system and the lower limbs, which undergo a growth spurt. The larynx also feels these effects (the voice breaks). Psychologically there is at the same time a withdrawing from the protective sphere provided by parents; the cutting of the umbilical cord is repeated at the psychosocial level. Hand in hand with this goes the awakening of the power of personal judgment that is turned very forcefully upon the world around it. The young person goes through an "identity crisis." Wrenched out of communion with a sensitive and sympathetic environment, he or she must now fulfill the role of critical outsider.

Apart from the sexual sphere, where development is complete, the Pygmy's physical maturity remains partial—the growth spurt in the legs is suppressed. What does this tell us? The relative length of the limbs is in a certain sense a measure of emancipation from the environment, of "inwardness"; for example, cold-blooded vertebrates have short extremities in comparison with warm-blooded ones (Bakker 1975), their body being raised only slightly from the ground, if at all. The contrast with mammals and birds is strong, and what is more these have their limbs beneath their bodies, not to the side like amphibians and reptiles. In their way mammals and birds attempt a kind of uprightness, or at any rate lift the body away from the ground. The body is also wrapped in a coat of fur or feathers, and the combined effect of thus insulating and setting the body apart is an expression of the animal's physiological autonomy in the matter of warmth, which in turn is the basis of an "inner life" independent of the momentary effects of ambient temperature and sunlight. Nevertheless, warm-blooded animals are still very much bound up with their environment through their instincts. They lack the psychological autonomy of human beings, which is not purely at the mercy of nature but has the potential for self-motivated action and bestows the ability to create culture. In both animal and human this distinctive feature of the degree of emancipation from the environment makes its morphological mark. In the former it appears in those parts of the body that come into direct bodily contact with the environment. The jaws and the ends of the limbs are not only the most morphologically refined and specialized parts of the body, but are also considerably elongated as a rule. In the case of humans, as has been shown by Schad (1977), we have precisely the opposite picture. The end parts of the limbs remain not only unspecialized, but also particularly short, while the upper leg, in contrast to the mammals, grows especially long. Thus while the animal restricts its impulse toward uprightness to a lengthening of the body parts closest to the ground, leaving its center of gravity out of the process and the head to droop earthward, the same process in humans renders the whole body, and not least the head, separate from the environment, against the effects of gravity. This process is not completed with the infant's learning to stand, for the growing legs continue to lift the body further aloft. The lengthening of the legs, chiefly of the femur, is the last phase of growth in the course of childhood development, most strongly evident in the well-known growth spurt around the time of puberty, and then fading out toward the age of twenty. In parallel, growing adolescents go through the process of "learning to stand on their own two feet," coming to rely on their own inner resources apart from the social environment.

If, as it would appear, there is a genuine correlation between physical and psychological distancing from the environment, then, since Pygmies do not go through this final surge of growth in their legs, we should expect that for them the psychological distancing would remain at most rudimentary. This does in fact seem to be the case. Among the peoples of their region they stand out as having distinct, childlike traits. Among these are their playfulness and love of physical movement, which can have them improvising dances on any pretext. Often it is the children who take the lead here, coaxing the adults with this or that gesture until they too join in. Many of these improvisations are regular "imitation dances," in which an elephant or chimpanzee will be portrayed with great dramatic skill and then tracked in the traditional way by other members of the tribe, and finally "killed." The imitative tendency shows too in the fact that in many areas the Pygmies have replaced their own language with that of their nearest neighbors (Maes 1951). Moreover, the strange symbiosis with the Bantus and Himas of their region, which has made the Bambutis economically very dependent upon them (Schebesta 1938), strongly resembles the child's relationship to parental authority.

Describing any particular people as childlike always tends to carry a negative connotation, implying that they are being judged to be below some norm of adulthood. This opinion could not be further from the truth! The single most fundamental characteristic distinguishing human beings from the mammals is not their highly differentiated brain, but the juvenile quality of many of their bodily features—the "fetal" proportions of the skull, especially the jaw section, the unspecialized hands, etc. (cf. Montagu 1984). These features lose something of their contour through the processes of growth, but nothing like to the extent that this occurs in apes, humankind's nearest relatives among the mammals. The infant is thus "more human" than the adult, quite simply because it possesses to a more marked degree the typically human features.

But to be a child also means having a different relationship to the world. In the child's world there are no "things" in the adult sense of external objects. The trees, the water, the stars, and the wind are all animate presences of the same nature as the child itself, and they are approached in this same spirit. Modern researchers such as Hallpike have shown that the animistic consciousness of so-called primitive peoples exactly corresponds to this way of relating to the world—there is not yet any alienation, no separation of subject and object, no emancipation. Nor is there any desire among hunter-gatherer cultures to change the natural order. They do not turn the phenomena around them into objects of their own intentions. This is a key feature of later cultures that have undergone a change enabling them to create a sphere of purely human activity, normally referred to as civilization. This process first turns human beings into farmers, then into city-dwellers, alienating them further from nature with every step.

Forest-dwelling peoples are just as powerless and helpless against the depredations of modern industrial cultures with their soulless, mechanized efficiency as children suddenly wrenched from the bosom of their family. Unlike agrarian cultures, which are astonishingly resilient, those still at the earlier hunter-gatherer stage, and therefore not used to interfering with their unspoiled natural environment, simply fall apart in the face of such challenges. The fate of the Bushmen, the Australian Aborigines, and many tribes of South American Indians is well known—among the last group, those first visited by missionaries were also the first to die out (Goodland and Irwin 1975).

For those at the agrarian stage of culture, however, the forest has always been the enemy, and not only in the tropics. They kept away from it—the range of the high cultures of the Andean Indians halted abruptly where the edge of the rainforest abutted on their mountain fields—or cleared it. In the tales and sagas of oral tradition, the forest has always been the dark maze, full of wild animals and other dangers, and as such corresponds to the dark side of the human unconscious that entangles people in its compulsive urges and dulls their mind (cf. Lenz 1971). The reckless destruction of the forest is still going on worldwide, and the smug remark of an Australian farmer to the effect that "one blade of grass is worth two trees" (Walter 1964) would appear to enjoy universal approval. This attitude, which was fully justified in the early stages of the agrarian push against the for- est, has now become a destructive power, which before long will have succeeded in laying waste what is one of the Earth's largest and most important ecosystems (Beck 1974; Gómez-Pompa et al 1973).

The cities of Europe, embodying as they do a culture more alien- ated from nature than any other, have recently seen (especially in Ger- many) an upsurge of nostalgic regard for the forest. While this might appear to be hopelessly romantic and unrealistic, born of ignorance in countries where the original forests have long vanished, there is surely something more behind it: namely, the sense of loss and inner poverty that is the legacy of alienation. In North America, on the other hand, things have been more direct. From the experience of the beauty and rich diversity of wild forests and of their being wantonly destroyed, the idea of conservation was born and first put into practice. Since then it has spread all over the world and today is based less upon romantic ideals than upon the insight that the future of the biosphere largely depends upon the survival of the great forest regions.

The Desert

Let us now turn to the opposite pole, the desert, and try to sketch a few of its typical features.

The pure, absolute desert is devoid of all life. Life exists only in the oases, and they are islands, living enclaves populated by domesticated plants and animals. Plant life is also present where rain falls regularly, even if at long intervals. At such places a form of growth known as the *succulent* has developed, which keeps its surface as small and imper- vious as possible in order to reduce to a minimum the loss through evaporation of the large quantities of water it takes in when the extremely intermittent rain falls. The prototype here is the globe- shaped cactus, which does not occur in Africa, although flowering plants with cactoid forms are found in the margins of African deserts. More typical, however, is the plant that, while keeping its upper parts

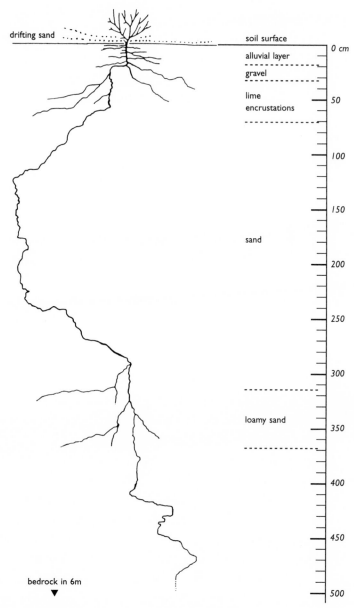

Fig. 7. Root system of the North African desert plant *Pituranthus*, which reaches down to the ground water (from Kausch, in Walter 1973).

very sparse, even living without leaves for long periods, has a root capable of boring all the way down to the deep water-bearing strata. But there are parts where the groundwater is inaccessible, and years go by without rain. That is the real desert. Through wide tracts of stony desert, or among the mighty combers of the wandering dunes of the sandy desert, it is possible to travel for days on end without encountering the slightest trace of plant life. Such areas are found in Libya, in southern Algeria, and in the "empty lands" northwest of Timbuktu. Astonishingly enough, in just such areas an addax antelope will turn up from time to time (Monod 1958). In this the animal, and particularly the *higher* animal, demonstrates the great advantage it has over plants in being mobile. This is an expression of its greater independence from its environment, its "inwardness" both physically and mentally. Higher animals have many more ways of avoiding the deathly clutches of the desert than plants. Nevertheless, confined as they are to one spot, plants do attempt, in their cactoid growth forms, to turn themselves inward.

To live in the desert means to persevere against it, to live in spite of it. The *animal* has a variety of ways of achieving this: it can hide away during the heat of the day and only come out at night (insects, rodents, fennec); it can develop the ability to reduce evaporation of body fluids to the barest minimum and so not need to drink, like the gazelle, certain antelopes, and the desert fox, which derive their water requirements from their food (Taylor 1969). A particularly amazing ability has made the dromedary the ship of the desert: because its blood is not affected and does not thicken even with acute water loss, it can tolerate a loss of up to 40% of its body fluids (25% of its body weight) without harm (a 10% reduction in body weight through water loss is fatal to humans). Sand grouse range far and wide for water and bring it to their young saturated into their underfeathers (George 1969, 1976). The sooty falcon (*Falco concolor*) delays its breeding time till autumn, when the flocks of birds migrating over the Sahara from the north provide a rich source of food for its young (Moreau 1966).

It goes without saying also that *humans* live not from the desert, but in spite of it, laboriously protecting their oases from the onslaught of the sand, or constantly moving with their hardy animals. Here, in contrast to the forest, everything contributes toward throwing human inhabitants back on their own inner resources, which they must draw upon or perish. Nature gives them nothing; everything they have they must wrest from it, and this gives them great strength. The early civilizations, particularly that of ancient Egypt, were oasis cultures that sprang up at the edge of the desert or semidesert. They did not "grow out of nature," but rather were products of human ingenuity added to nature. The Egyptian culture flourished increasingly as the last green islands in the Sahara, remnants of the lush vegetation that covered the

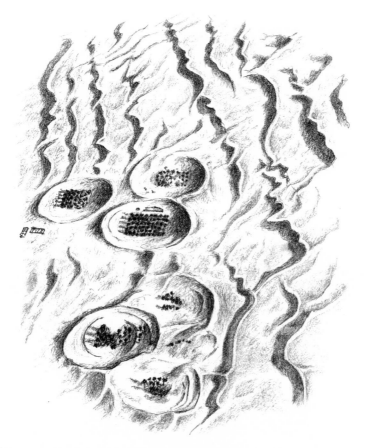

Fig. 8. Palm oases in the Souf region of the Algerian Sahara, drawn from an aerial photograph. The photographer Georg Gerster writes: "Here this waterless ocean is laced with waves of high dunes. Thousands of craters wrested from the sand by staggered rows of palm leaves harbor large and small groups of date palms, their crowns often hidden behind the defensive walls. . . . Because the water did not come to them, the Souf dwellers, together with their palms, sought out the water: At the base of these agricultural bowls, which vary in shape, size, and depth, they planted trees so that their roots tapped into the moisture collecting beneath the sand. This arrangement offers a classic example of successful human opposition to unfavorable natural conditions. However, it seems that the Souf desert will have the last say. Oil riches are threatening the date economy: young men, lured to the new oil fields, are finding the walk to the gardens too tedious, not to mention the drudgery of maintaining them. Crater after crater is reclaimed by the sand." (1975)

area during and after the last Ice Age, died away. The Egyptian temple "is imposed . . . on the landscape." It spurns the outside world, in marked contrast to the Greek temple, which is never incongruous, but crowns the landscape in which it stands (Gut 1974). The Kasbahs and Qsare of today's desert-dwellers are also like siege-proof fortresses.

Another way of living in the desert is exemplified by the nomadic existence of the Bedouins, whose livelihood depends upon very widely spaced wells, herds, and the availability of grazing. They embody in its purest form a culture of self-reliance, austerely and proudly masculine, motivated by aggressive self-interest directed toward the subjugation of enemies, oasis-dwellers, slaves. Raids upon oases have long had their established place in this culture. In social terms the Bedouin are shep-

Fig. 9. The Kasbah of Tansirt near Zagora at the foot of the Anti-Atlas mountains in the Moroccan Sahara (above). A Pygmy hut of banana leaves (below). It is hard to imagine a greater contrast than between the defiant, fortresslike buildings of these desert people and the forest-dwellers' frail rain shelter, which hardly separates them from their environment.

herds and warriors, the polar opposite of the peaceful, childlike Pygmies and the black farming peoples of savanna and forest, against whom they conducted slave raids for many centuries. While it is true that the Bedouins stem from Asia Minor, the fact remains that the Arabian Desert, right down to the animals and plants that live in it, is simply the eastward continuation of the Sahara across the Red Sea.

If the lifestyle of the Bedouins (and Tuareg, as well as other nomads) demonstrates one side, the cultures of the oases represent the other. What characterizes both in the same way is *life in spite of, or against, the desert.* The magnificent systems of water management and use— the blooming, overflowing, artfully enhanced gardens, waving fields of grain, even shadowy forests (of date palms)—conjure up a completely artificial, "human-made" nature, with a razor-sharp border at the desert where no further water can be brought. We can only admire the rich inventiveness with which water is led from distant regions, with no loss through evaporation, as in the subterranean canals of the Qanat in the Iranian desert, or in the refined systems developed by the Nabat for reception and retention of winter rainfall in the rocky wadis of the Negev.

It even appears that our modern technological/natural-scientific culture derives from the constant compulsion to develop a technical, engineering intelligence in the struggle with a life-threatening nature. In this regard, the Arab-islamic oasis cities of the Middle Ages were already considerably further along, more "modern" than Europe. Through the incursion of Islam, Europe received a developmental shock that continues undiminished to this day, and in fact is constantly increasing. Clearly, this is connected to another phenomenon, which proves increasingly dangerous in a less life-threatening environment: the tendency to concentrate exclusively on the construction and expansion of human-made, increasingly artificial "cultural oases"—residential developments and highly unnatural agricultural lands—while misunderstanding and pushing back the surrounding natural landscape as "wilderness" or "wasteland." One reaction to this is the growth of environmental movements precisely in those countries that belong to the principal proponents of this development.

If the *primeval forest* is the high point of vegetative life on Earth, a climax so one-sided that higher animal life retreats before it, human culture is prevented from unfolding, and minerals are completely integrated into living processes, so the *desert* is the place where lifeless stone in all its various forms, from finest sand to giant bastions of rock, holds absolute sway. It is quite literally a landscape that has died. The traces of long spent living processes are still clearly to be seen, just as the imprints of muscles and blood vessels remain on a fossil bone. These are the ancient riverbeds, the ravines carved out by post-glacial waters, and the rock drawings of human figures as shepherds and hunters and of the rich animal world that now inhabits the tropical savanna and includes forms from the north. The very last evidence of this earlier presence of life is found in a small number of mighty cypresses still growing in the desert range of Tassili N'Ajjer. Some of these have been aged at 4,700 years old—they germinated as Cheops was building his pyramids. Under the present extremely arid conditions new growth does not occur, even though the seeds prove viable when planted in botanical gardens (Knapp 1973; pictures of *Cupressus dupreziana* in Gardi 1970). Moreover, pollen of spruce (*Picea orientalis*) and yew (*Taxus*) from the last Ice Age has been found 2200 m up in a range of mountains in the southern Sahara (Walter 1973).

There is even an ancient Egyptian chronicle that pictorially records the gradual advance of the desert; its animal illustrations document the striking changes in fauna that took place over the millennia. In pre-dynastic times, between 4000 and 3000 B.C., the most commonly depicted animals are elephants and giraffes, in addition to lions, African buffalo, black and white rhinoceros, and even aurochs and fallow deer. With the transition to the Old Dynasty the elephants, giraffes, and rhinoceros give way to antelopes such as addax and oryx, and other gazelle-like species of arid landscapes—a tendency that increases with the advent of the Middle Dynasty (Butzer 1959).

Water, once the bringer of life and fertility from the atmosphere, has now retreated deep under the earth. Like a fossil it lies under the desert mantle, thus embodying the counterimage of the rainforest, which is continually clothed in a thick layer of moisture but stands in impoverished and demineralized soil. What the one landscape has in excess, the other one lacks.

The Sahara is a landscape that, in its current lifeless state, is entirely opposite to the way it was formerly. Whereas today it is the gigantic no-man's-land between black Africa and the Middle East and Mediterranean regions, blocking the spread of all living creatures, in former times the very same tracts of land were the bridges over which ostriches, elephants, and many other species made their way as far as Morocco and Palestine. In the other direction members of northern plant genera penetrated deep into the south of the African continent, as we have already observed in our consideration of the upland forests. The locations of the remaining stands indicate the pathways by which this occurred (Fig. 10).

Fig. 10. A. Range of the elephant shrew *Elephantulus*, a species limited to Africa alone (after Kingdon 1974). As a result of the desertification of the Sahara, the range of many African animals is similarly dispersed. **B.** Range of the African pencil cedar (juniper) tree *Juniperus procera*. A young and old tree are shown at right. In Arabia these trees occur together with closely related, smaller species that have spread from the Mediterranean: *Juniperus excelsa, foetidissima* and *phoenicea* (not indicated). The Red Sea does not present a barrier to expansion. Due to sinking sea levels at its southern end during the Ice Ages, land bridges repeatedly connected Arabia and Ethiopia. (Elevations above 1000 meters are dotted. Map combined after Moreau 1966, Knapp 1973, and Walter 1973.)

Our impressions of the desert are summed up in a succinct and evocative description by the great Sahara explorer, Gustav Nachtigal: "The gently rising plain before me, covered with small, smoothed pebbles, outdid all the plains I had so far seen in the magnificence of its uniformity. Nothing was there on which the eye could find purchase, not even the merest trace of life. A complete image of emptiness and infinity. Nowhere does the human being feel so small and forlorn, and yet nowhere so strong and exalted, as in the struggle with the pitiless desolation of this lifeless, seemingly unbounded fastness. Traveling in the desert makes him grave and brooding, and the true sons of the desert, the Tauregs and the Tubu, who spend their whole lives in this lonely fight against the vast wasteland, have an almost sinister set to their features, which it seems no innocent mirth could ever again lighten."

The Savanna

If the desert can only treat us to vast, monotonous, unknowable spaces and the rainforest to close, and in its way equally uniform, confinement, the savanna, for its part, encompasses both qualities. Viewing it from the top of one of the many, small extinct volcanoes scattered over the East African plateau one sees a mottled carpet stretching over the earth to the horizon: an ocean of round, unevenly spaced, green spots —acacias, as far as the eye can see. On closer inspection, however, the uniform picture dissolves into a variety of small landscapes. Every tree has its own particular shape and stands singled out above the undergrowth, which glows either bright green or yellowish brown according to the season. The huge figures of baobab trees also stand here and there like giants gesticulating at the sky. At some places the trees thin out, making way for stunted clumps of whistling thorn and small, extremely prickly acacias, before yielding to pure grassland. The small trees sport black, gall-like growths in which ants live. The stands of parasol-like acacias may also be interspersed with artistically spaced groups of slender Borassus palms, which congregate in dips and hollows where rainwater lies longer. Like thin green snakes the narrow

Fig. 11. The striking contrast between the umbrella acacia and the baobab (the African savannah's two characteristic trees) comes clearly into view in the dry season. While the acacia, under its thickly leaved crown, offers animals urgently needed shade from the burning midday heat, the baobab stretches a tangled mass of leafless branches into the sky. In the former, the crown is the main event, while the trunk appears weak and powerless; in the latter, it is exactly the reverse: the powerful trunk dominates the visual phenomenon of the tree. It has particularly soft wood that can store vast quantities of water during the dry season. Elephants tear out great chunks of it with their tusks to suck on. (Local tribes know very well how the baobab got its unusual appearance. When God the Father created the trees and placed them on Earth, the baobab alone was unhappy with its placement. When it was transplanted, the new location turned out to be just as unacceptable, and so it went everywhere God put it. Finally, God lost all patiennce, grabbed the baobab, tore it from the earth, and planted it again upside down. Since then, it has stretched its roots, in bewilderment, up into the sky.)

bands of tangled gallery forest wind along the watercourses, dominated by large, dense-leaved sycamores.

As soon as we come down from our high viewing point the landscape is transformed as if by magic. We suddenly find ourselves on what seems like an island of meadow loosely set with trees, a spacious clearing surrounded by forest. But if we try to reach this forest it retreats before us; and no matter how far we walk we could never reach it, for it proves to be a trick of perspective—the further away the trees are the closer together they appear. Now we begin to notice the abundant animal life. The trails of grazing animals wind through the grass in all directions like rivulets. Mixed groups of zebra and wildebeest amble purposefully past, lie at their ease, or gently browse. A small herd of giraffe literally materializes before our eyes by stepping out of the dappled shade of a wide-topped acacia that had rendered them completely invisible. A strange flowing rustle in the grass betrays the presence of an African rock python, which glides down the hill glinting in the sun like liquid metal. On the ventilation tower of a brick-red termite nest a tawny eagle perches. A tribe of baboons takes flight before us to the shrill cries of the females and children and the angry yapping of the males. Alarmed, a herd of impala spring gracefully away in great leaps, each one like an arrow shot from a bow. Metallic blue flashes of glossy starlings with their bright red bellies scatter with complaining cries to the treetops where they join the black and yellow weaverbirds, whose gourd-shaped nests sway like dried fruits in the branches.

Among a large herd of grazing animals sedately crossing the scene in the distance there are suddenly some flashes of silver. The binoculars soon tell us that they were from the metal heads of spears belonging to Masai herdsmen tending their cattle. They ply their trade among the wild animals, which scarcely move aside as they pass. Over some trees stand strange columns of smoke with flocks of birds circling around them. On coming nearer we see that they are swarms of termites. By the million they have risen on temporary wings for their mating flight, only to fall victim to the waiting predators of the air— white-tailed African house swifts, and among them large, sharp-contoured Alpine swifts, winter guests from Europe. In the midst of such overwhelming abundance it is a problem knowing where to look first, so that it is all too easy to overlook what is growing at one's feet—the great, scarlet, turban-shaped blossoms of the *Gloriosa* lily, which twines around grass blades with the tips of its leaves, or the giant, flame-red tufts of *Haemanthus*, whose round flower heads consist of innumerable star-shaped blossoms. Jet-black scarab beetles are eagerly rolling their balls of dung to safety before a rival can steal them. And time after time the great shadows sweep by of vultures sailing low over grass and trees with no hint of a wing-beat, always on the lookout for carrion.

Returning to the scene a few months later, one finds that all the fullness and color have disappeared. Pale brown grass crunches underfoot, and large locusts whirr off into the sky. The great herds have moved on, the nests of the weaverbirds are now empty straw balls swinging ruffled in the wind, the vultures have vanished from the sky—there is not a scrap left for them. A solitary secretary bird hurries by paying us no heed, intent upon its business of stalking snakes. The smell of burning hangs in the air. Time and again we come upon blackened areas where the charred trunks of the acacias make a very desolate picture.

From our observations four features peculiar to the savanna stand out particularly clearly:

1. *the merging of grassland and woodland;*

2. *the abundance of birds and mammals;*

3. *seasonal change and its influence on the animal and plant worlds;*

4. *the presence of pastoral nomads.*

In the savanna, as *arboreal grassland,* two vegetation types that normally exclude each other in their ecological requirements have formed a harmonious unity (cf. Walter 1968, 1973, upon which the following is largely based). In the grasslands of temperate latitudes, the Eurasian steppes, and the North American prairies, there are no trees, since the dry season comes during the peak time for vegetative growth, in summer. Admittedly "forest steppe" does exist, but only as a mosaic of islands of closed forest, growing in moist hollows and along watercourses amid unbroken grassland. A true interpenetration of steppe and woodland does not occur, and where it is artificially attempted, for instance for the purposes of establishing windbreaks, the trees quickly wither and die. For grasses, however, the conditions are ideal, since they can use the short period in spring when the ground is wet to go through their whole life cycle. During the dry summer they become dormant. Their next year's growing tips are safely tucked under the withered leaf blades, and the outer layer on the roots dies back to form a protective sheath around the still living central capillary tube. In the regions north of the steppe, for example in Russia, no grassland can form, for there the summers are wet and favor the growth of trees, which keep light from reaching the ground. In such shady conditions no grass can flourish.

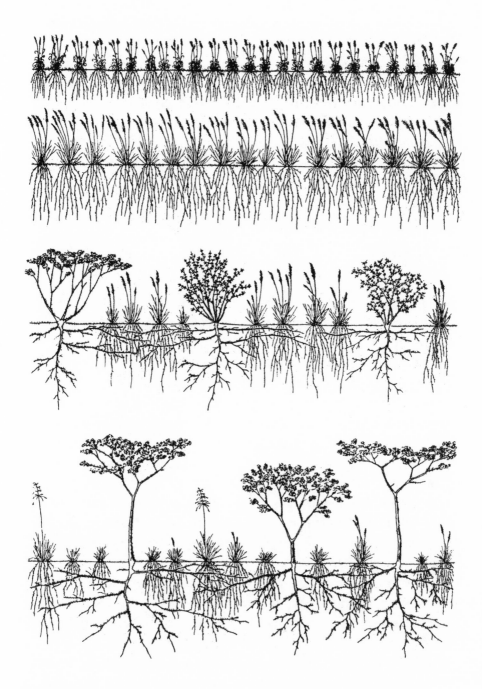

Fig. 12. Savanna between grassland (above) and open dry forest (below) (from Walter 1973).

Of course, this applies only to natural plant communities and not to landscapes created by human cultivation. In temperate regions trees formerly dominated the scene. The agricultural landscape with its varied patchwork of fields, meadows, woods, and copses gradually took shape throughout the Middle Ages as the large-scale clearing of the forest proceeded. Thus a landscape emerged that has much in common with the savanna. Botanists exploring East Africa at the beginning of the twentieth century felt constrained by this extraordinary similarity in landscape physiognomy to name a certain formation in the savanna "orchard steppe" (Engler 1910). Of course, this type of savanna is, it should be remembered, an entirely natural phenomenon, whereas the European orchard is a cultural artifact that would very quickly revert to forest if left untended by human hands. Nonetheless, this correspondence is by no means exhausted in a purely pictorial resemblance. The savanna also has certain "hands" tending it in the form of grazing animals and fire. We will return to this important fact later.

Grass and trees coexist in the savanna because just enough water gets into the ground in the rainy season to keep the trees going through periods of drought, assuming, of course, that their numbers are restricted and they are thrifty with water. The latter they can accomplish either by shedding their leaves in the drought season or by keeping their surface area so small that transpiration is reduced to a minimum. The acacia, with its small leaves that are divided into many thin leaflets and fold up in the heat of the day, fulfills this demand very well. Its thick bark and tough bud casings tend in the same direction, and are features that it has in common with temperate deciduous trees, which treat the winter as a time of drought.

The loosely spaced trees offer no competition for sunlight to the grass, which is thus able to grow abundantly, nor do trees and grass have to compete for water since their ways of obtaining it are so fundamentally different. The trees of the savanna have "extensive" root systems, which means that they radiate out widely and reach down deeply—to a depth of forty-five meters in *Acacia giraffae* and as much as fifty meters in *A. tortilis*—and are therefore able to reach pockets of moisture even at times when the densely tangled "intensive" roots of the grasses have sucked the ground's upper layers dry (and the grasses have long since dried up). That only a few trees are able to exist and that they are far apart is nevertheless understandable. Where the precipitation rate increases and correspondingly more moisture remains in the ground during drought periods, the trees multiply: the savanna becomes dry forest, and the grass fails for lack of light. But where the rainy season is not sufficient to sustain moisture in the ground, the trees cannot survive despite their deep roots: the savanna becomes grassland, which with increasing aridity gives way to the thornbushes of the semidesert, and eventually to pure desert.

These transitions can be observed with particular clarity in the region around the giant crater and lake of Eyasi in northern Tanzania. The western slopes of the craters Oldeani, Makarut, Ngorongoro, and Olmoti are low in precipitation and therefore dry. Like the tracts of the eastern Serengeti into which they merge, they are covered with treeless grassland. They lie on the dry side of the watershed, for the trade winds off the Indian Ocean deposit their rains on the eastern slopes. Moreover, these eastern slopes of the craters are clothed in extensive cloud forests, while fairly thickly wooded dry savanna spreads over the lower slopes. Going through this landscape in a southerly direction, paced to the west by the gradually diminishing wall of the crater, we find that the wooded savanna slowly turns into thornbush savanna and then, on the flat, into semidesert. As the acacias become less frequent the giant water-storing trunks of the baobabs come to dominate the scene. Small, stunted thorn trees of the species-rich genus *Commiphora* (myrrh) now proliferate, joined by many kinds of succulents, which bear witness to the fact that the plant world has, as it were, withdrawn into itself in the face of the unfavorable living conditions. There are treelike euphorbia (*Euphorbia nyikae*) beside other small *Euphorbia* species, looking for all the world like erratically spiked cacti, and *Caralluma* species with pallid, gray-green, fleshy stems topped by the dark globes of their deep-purple blossoms, which emit a revolting carrion smell. Also cactus-like is the liana *Cissus quadrangularis*, which can engulf whole trees and is reminiscent of

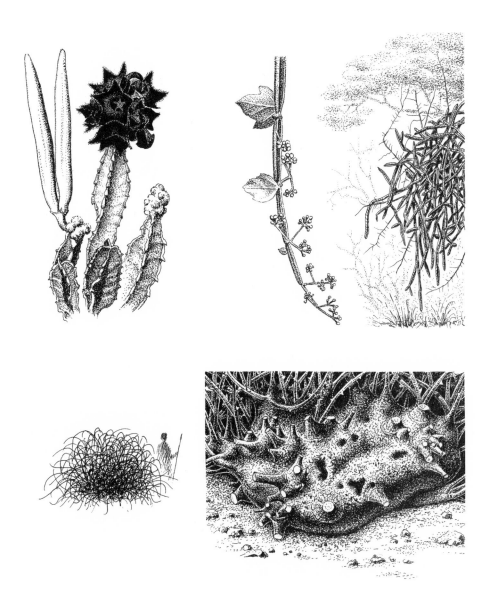

Fig. 13. Succulent plants of the dry thorn forest and semidesert near Lake Eyasi in northern Tanzania. Above left: Atop its fleshy, cruciform—and edible—stem, the black-violet inflorescence of a large *Caralluma* exudes a strong smell of decay. Above right: Bushes and trees are overgrown with the segmented sprays of a species of *Cissus*. Above center: During the rainy season blossoms and leaves appear temporarily on the new shoots of this cactus-like plant, revealing its kinship with the grape (*Vitis*). Below: "Elephant foot" (*Adenia globosa*). When the wild, thorny stems are removed, the juicy green rind of its soft, watery "trunk" is exposed.

the twining *Cereus*. Only in the rainy season does it put out its small, bright-green leaves and the yellowish flower clusters that betray its close kinship to the vine (*Vitis*). Thrusting up between the rock fragments of a weathered outcrop are the bottle-shaped stems of an (apparently still never described) *Adenium,* which in the dry season unfolds delicate pink blossoms reminiscent of oleander. Then there is "elephant's foot," uniquely grotesque but still rather hard to find among the thornbushes. It lies hidden under a dangerously thorny tangle of whiplike rods raying out every which way. These must be chopped away with a bush knife to get at the peculiar trunk, which looks like a large, knobby, bright-green stone (*Adenia globosa,* Passifloraceae). But this stone contains plenty of moisture, so that the plant can flourish even when the air is so fiery hot and conditions so dry that our lips split. Not two hundred meters further on is a broad patch of shade under the green domes of some tall sycamores, their roots bathing in the clear water of a spring surrounded by green clumps of papyrus. This miracle is due to the spring, which does not dry up even at the height of the dry season. Wherever there is a plentiful supply of water, the forest moves in, usually in the form of gallery forest, threading its way through the surrounding aridity and composed mostly of yellow-trunked fever trees (*Acacia xanthophloea*) and sycamores (*Ficus sycamorus*).

More than any other, East Africa is the region that sets before us the whole variety of African landscapes, either sharply juxtaposed or intermingled. The savanna belt thus consists of a mosaic composed of desert-like tracts, small areas of rainforest, savanna, and thorny scrub: the whole of Africa in miniature. This is particularly true of the areas around the two imposing, solitary figures of Kilimanjaro and Mount Kenya. The picture is completed by the presence of all three of the main tribal groups of Africa—the Bantu, the Hamite, and the Bushman—who also intermingle in mosaic fashion like the different landscape types.

No continent possesses such an abundance of large mammals as Africa, whether in terms of numbers of species or of individuals. The majority of them live in the savanna. From the open dry forest to the semidesert there are sixty-eight species of ungulates alone (Bourlière 1973), as opposed to six in the corresponding landscapes in South America. There are a further nine desert-dwelling species (six gazelles, two antelopes, and the Barbary sheep). Of the twenty-seven pure forest-dwellers the majority fall within a size range between hare and small deer (royal antelope, duiker, bushbuck); only three attain the proportions of the European red deer, namely the okapi, bongo, and mountain nyala. In the savanna, however, it is the large ungulates—eland, wildebeest, oryx, giraffe, and buffalo—that dominate the animal scene far and wide. And the sheer numbers of beasts on the move, so typical of the savanna, are also something unknown to the forest: the teeming multitudes of wildebeest and zebra, the springbok that used to roam in the hundreds of thousands, etcetera. Whereas the forest has animals that hide themselves from view—consider the case of the okapi, which eluded discovery until 1936—leaving the scene totally dominated by the plant world, and the desert is ruled by the dead world of minerals, the savanna would not be complete without its roaming herds, both large and small, its elephants and rhinos standing like blocks of stone on the plain, the colorful glory of its bird life.

Plant life and animal life in the savanna are very closely attuned to each other. The species that occur in large herds are all pure grazers and prefer—or preferred—wide open spaces with few trees, as found in South Africa or the Serengeti in East Africa (for example, hartebeest, topi, both kinds of wildebeest, Ugandan cob, springbok, and buffalo). This large-scale association of grass and grazers did recur in northern latitudes but with one major difference: the giant herds were always formed by only *one* species of animal. The North American prairies had their bison, the southern Russian steppe its saiga antelopes. The undisturbed African savanna, however, always presented a picture of a large variety of grazing animals in groups of varying size in which species either mingled peacefully or lived as very close neighbors: "In the more remote parts of South West Africa one usually finds several species of antelope grazing in close proximity to giraffes, zebras, warthogs or ostriches. . . . The tolerant, not to say downright friendly, ways African ungulates often have of relating to each other are most strikingly expressed in the wildebeest's behavior toward other species. Where large game are numerous it is usual to see zebra and wildebeest, springbok and other antelope sharing the same ground. The wildebeest are especially sociable creatures. They not only run in large herds, where they are protected from attack, but often associate with other animals, especially Burchell's zebras, to whom they seem so friendly disposed that they are commonly found grazing together." On one occasion this author even saw an orphaned zebra foal being suckled by a wildebeest that had lost its own calf (Shortridge 1934).

The browsing species (giraffe, greater and lesser kudu, eland, nyala, etc.) tend as a rule to form much smaller herds and family groups, perhaps with the singular exception of the elephant. These then form the transition to the solitary forest-dwelling species. Here once again the tremendous ecological range of the savanna is reflected, which unites within its broad span forms with a preference for dry, even desert, conditions (gazelle, oryx) and forest-dwelling species (duiker).

With a bit of luck it is possible to have almost the whole span of diversity in view at once. The ideal spot for this is the Ngurdoto Crater at the foot of Mount Meru in northern Tanzania, and the ideal time to go there is the dry season, for this is a region that does not dry out. From the lip of the steeply sloping crater in late afternoon in the dry season the eye is treated to a panorama of the animal world of Africa, gathered around the numerous water holes or roaming through the extensive marshlands. Two female rhinos come with their calves to drink; behind them a third ambles toward another water hole. One of the babies is so small and roundheaded that it can be only a few days old. Groups of black buffaloes graze, lie chewing the cud, or just stand around vacantly. A long line of elephant cows with their young comes along, marching leisurely, as if in slow motion, straight across the marsh. A very small member of the troop, still at the mercy of all sorts of distractions, loses touch with the line and is left floundering around in the mud. After the herd has carried on for half a mile the mother notices the gap, turns round, and fetches the little dawdler. Away in the distance a party of giraffe crosses the dry meadows; one of them lies down in the grass. Right below us one bushbuck after another is coming out of the forest, eventually seven in all, dark rust-brown with black and white spots and sharp, tightly twisted horns; there may be a larger number of the paler females. They spread out grazing over the meadows along the forest margins. Warthogs shuffle around on bent forelegs digging for roots and paying no heed to the noisy, cantankerous baboons gamboling around them. Two red duiker, with the arched backs and drooping heads typical of species that take cover in thickets, emerge from the forest, the one hard on the heels of the other as they rush to cross the clearing as quickly as possible . . .

The large animals, of course, are not just incidental components of the savanna landscape. Some observers indeed go so far as to view them as the chief cause of the savanna's very existence (Knapp 1973). Proponents of this view point to Madagascar, which has no large mammals and where grasslands were unknown before colonization. The prevalence of thorny trees and bushes is also ascribed to the influence of browsers, which are said to have gradually driven back the "unarmed" tree species. Here Australia is offered as evidence, since it has savanna devoid both of thorny plants and of large browsing animals (the giant kangaroos eat grass). Both these arguments, in seeking to explain everything in terms of one cause, miss the mark. The existence of the savanna, as was shown earlier on the basis of Heinrich Walter's work, is at the very least a product of climatic factors. And plants armed with thorns are by no means free from browsing, but are cropped by ruminants highly adapted to the purpose (giraffes, gerenuks).

The causal picture is, as always with living processes, multifaceted. A host of different effects act upon, and counteract, each other to form a complex web of mutual dependencies—in other words, an organism—as is the case in any ecological community. In the functioning of this particular organism the role the herbivores play is decisive: to a considerable extent *they are involved in the making of the landscape*. In recent years the destruction of trees in areas of both dry and moist savanna in the Tsavo and Murchinson Falls National Parks has excited much attention. A complete change in both the landscape profile and the ecological balance has been the consequence (Laws 1970), and the elephant has been identified as the prime cause. Thus its ability to transform the landscape is not in doubt. The question is, however, whether the change came about naturally, or whether it was not due indirectly to human interference. In the areas outside the parks the elephants are constantly being hunted, so that they take refuge in the protected areas, where so many may accumulate that overpopulation is the result. This must then be controlled by culling. A phenomenon such as this makes the shortcomings of national parks apparent. They tend to end up as isolated enclaves among lands ever more densely settled and intensely cultivated, cut off from living interaction with their surroundings. A further consequence is the increase in poaching, which with growth in human population density and technical "progress" takes on ever more extreme forms. This is all perfectly understandable, for the people see no reason why they should not tap the rich game herds for food.

It has in fact been demonstrated that the savanna can support much higher numbers of wild than of domesticated animals. In healthy savanna in its natural state, the biomass of ungulates found is significantly higher than what, in the form of cattle, will result in overgrazing to the point of soil destruction. In arid areas, such as the Sahel, this overgrazing will lead to the formation of desert, while in other zones it will produce thorny, impenetrable bush (Walter 1964). One answer to this situation has been repeatedly recognized and called for: sensible wildlife management, in other words the commercial husbanding of wild populations instead of their elimination. We will return to this point later.

Wild grazing animals, unlike their domesticated counterparts, are not confined in their wanderings to a narrow range, but roam over large areas. Since they never linger very long in one place, the vegetation always has a chance to recover from their heavy grazing. The grazing also occurs in a definite sequence: "The first on the scene are the zebras; over a few days thousands come past . . . on their way north, cropping the grass that has shot up in the rainy season. Then come the wildebeest . . . strewn out in long herds of up to 10,000 moving northward day and night, and the air is alive with their lowing and rife with their smell. . . . After a few days the grass has been nibbled right down to below ankle-height. The wildebeest gone, the plain is suddenly filled with lively, tail-wagging Thomson's gazelles, which usually remain some weeks in the area" (Hendrichs 1971). Behind this sequence lie different food preferences that mutually enhance each other. The zebras favor fresh green grass still in bloom; the wildebeest prefer grasses that the zebra do not touch, while the gazelles eat the leftover stubble (Kingdon 1971). Savanna that has been so fully grazed provides much less fuel for the fires of the dry season than does an area of dried, uncropped grass. Here there is no chance of a fire so hot that it destroys soil fauna and grass roots, and kills acacias otherwise protected from fire by their thick bark.

There are also ecological relationships of a more intimate nature than those just sketched; just how intimate is shown by the connection between acacias and antelopes. *Acacia tortilis* sheds its seedpods during the drought, when the grass has dried up, been eaten, or burned. Antelopes eat the pods for their sweet and nutritious shells, depositing the seeds along their trails and at their resting places. These seeds have a higher germinative capacity than those that have not been through the digestive tract of an antelope. What is more, they remain protected from attack by beetles, which burrow into the uneaten acacia seeds. Apart from having food during a lean time, antelopes profit from this in another way, especially eland. They must have shade during the midday heat, and where better to find it than under an acacia "parasol" (Kingdon 1971).

In many areas (especially those where there are large concentrations of grazers like the Ngorongoro Crater and the Serengeti) the *beasts of prey* cut a significant figure—the lions, cheetahs, hyenas, jackals, and hunting dogs. That they are quantitatively negligible—in the Serengeti small and large predators make up only 0.6% of the total mammal biomass—need not belie their real significance, for this is typical of all predator-prey relationships among warm-blooded animals (Bakker 1975). For George Schaller (1967, 1972), who has wide field experience in Africa and Asia, the significance of predators lies in the fact that they keep the grazer population below a certain limit beyond which hunger, sickness, and other factors associated with over-grazing set in, leading to drastic population fluctuations in which an initial explosion is followed by inevitable crash. Compared to such catastrophic alternatives, the selective and regulatory activity of preda-tors is much more discreet and conducive to effecting and preserving ecological stability.

The *birds of prey* may be viewed in the same light. Evidence for this is immediately apparent in the ecological separation that can be observed between similar-sized species sharing the same range. The consequences—for both the hunted and the hunter—would most likely be fairly drastic if all three species of eagle involved (Fig. 14) specialized in attacking the same prey. There is a certain amount of overlapping, "especially with regard to rock dassies (hyrax), but this . . . does not lead to competition, for the three typical species range over different habitats. The black eagle lives in rocky hills, the martial eagle in the open plains and the crowned eagle in the forest. Dassies occur in all three of these habitats wherever there are rocks, but their numbers vary from landscape to landscape. They are the main food source of the black eagle, which does its hunting where they are most numerous. The martial eagle will occasionally take a dassie if it comes upon stone outcrops in the open plain, but it rarely scrutinizes the hills as closely as the black eagle. The crowned eagle normally carries off what dassies it eats from the forest, where they often nose around during the day, concealed from the eyes of the black eagle by the for-est canopy. Thus it is possible to find these large and powerful birds living in close proximity. All three species are known to have bred on one hill in the Embu District (Kenya) without poaching on each other's preserves" (Brown 1970).

It is not so easy to perceive the role of the other *birds*. In the savanna, birds, with their calls and songs and the splendor of their colors, do every bit as much to animate the landscape with vibrant life as the mammals do. In comparison to temperate latitudes, indeed, both appear to have attained a higher level of life-intensity, which to the unaccustomed visitor is quite overwhelming. Take, for example, the sparrow family (Ploceidae); the rather drab coats of our few native

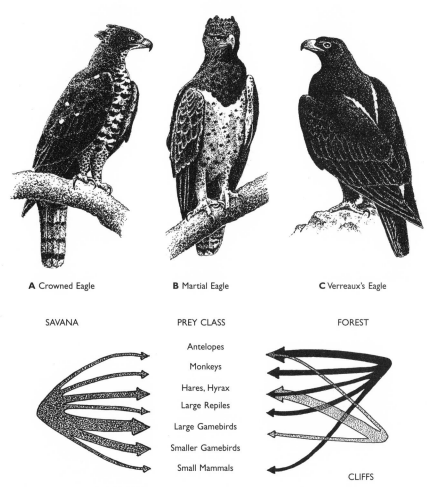

A Crowned Eagle   **B** Martial Eagle   **C** Verreaux's Eagle

SAVANA   PREY CLASS   FOREST

Antelopes

Monkeys

Hares, Hyrax

Large Repiles

Large Gamebirds

Smaller Gamebirds

Small Mammals

CLIFFS

Fig. 14. Ecological separation of three eagle species of approximately equal size that coexist in many regions of the African savanna: crowned eagle *Stephanoaetus coronatus*; martial eagle *Polemaetus bellicosus;* Verreaux's or black eagle *Aquila verreauxi*. (Diagram after Brown 1970, redrawn.)

species (for example the ubiquitous house sparrow) are transformed into the brightly colored markings—golden yellow and black or reddish-brown—of the innumerable African species of weaverbirds. In place of the untidy heaps that pass for nests among our sparrows (and also among certain closely related species in Africa) the weavers hang skillfully woven baskets from the tips of branches, a sight one cannot imagine the landscape without. The radiant colors of incandescent kingfishers, bee-eaters, glossy starlings, and sunbirds delight the eye

Fig. 15. Representatives of typical savanna bird groups; only the black-throated honeyguide *Indicator indicator* (above left, brown and white) is a forest species. Above center: Van der Decken's hornbill *Tockus deckeni*. Above right: Mousebird *Colius sp.* (brown, very long tail). Below left: Double-toothed barbet *Lybius bidentatus* (red below, black above, white flank feathers). Below center: Button quail *Turnix sylvativa*, in spite of its similarity, not related to the chicken but a representative of the independent family of the Turnicidae. Below right: Lilac-breasted roller *Coracias caudata* (azure blue and lilac).

and contrast sharply with the tawny background of the savanna. Well-nigh every birdcall imaginable is there, from scraping, tapping, and croaking to far-carrying whistles and the most marvelous songs, which in beauty of tone, melodiousness, and variation outstrip the best European songsters. The thickets of the savanna are home to a wide range of close relatives of the blackbird and nightingale. Birdsong is taken to new heights here, especially by the habit some species have of singing in duets (Thorpe 1973). In the case of the tropical boubou (*Laniarius aethiopicus*), for instance, the only way of telling that the perfectly fluid melody is sung by a male and a female is to keep a singing pair in view. (A good survey of the diversity of African bird songs can be had from recordings made by North 1958, and North and McChesney 1964.)

Like the mammals, the world of birds is not concentrated in the forest, but in the savanna: against thirty-nine families of birds consisting of species that live mostly in the savanna, there are only eight composed of those that spend most of their time in the forest. Only four families consist of exclusively forest-dwelling members, and only one of these is rich in species (the woodpeckers, which are found over the whole world wherever there is forest, except in Australia). The other three contain only a few species and are more richly represented in other continents (trogons: two species, widest distribution in South and Central America, India, and Malaysia; broadbills: three species; Pittas: two species, distributed more widely in the South Asian and Indo-Pacific regions). Apart from its two Asian species, the single purely African family of typical forest birds is that of the honey guides, close relatives of the woodpeckers. By contrast, there are fifteen families that live exclusively in the savanna, most of which are not confined to Africa, but are nevertheless very characteristic of this continent (vultures, falcons, secretary birds, storks, bustards, plovers, sandgrouse, ostriches, button quails, mousebirds, rollers, larks, pipits, cape sugarbirds, spotted creepers). Eleven other families contain more or less equal numbers of forest and savanna species, for example turakos, hornbills, barbets, pigeons, and the greater part of the songbirds (compiled from Moreau 1966).

Such a list is rather dry, but it does make clear the dominant role of the savanna in connection with the bird life of Africa. The position is strongly reinforced by a comparison with South America, where most families, among them those richest in species, are pure forest-dwellers. In a small area of rainforest in eastern Peru 408 bird species were counted (of which at least 300 were local breeders), whereas the whole African rainforest contains only about 250—as opposed to 855 in the open landscape (not including waterbirds; after Amadon 1973, O'Neill and Pearson 1974, Moreau 1966). Moreover, the numbers of birds in the African savanna are swollen each year by the flocks of

European migrants, of which, again in contrast to South America, only a small fraction make for the forest. The majority stay in the savanna belt between the Sahara and the equator, in other words the zone that has only one annual rainy season, which coincides with the northern summer. The migratory birds thus spend their winter rest together with the birds native to the region at a time when the drought has reached its height and food reserves their low point (Moreau 1972)— at first glance a somewhat incomprehensible biological phenomenon.

When viewed, however, within the larger context of the animal world of the savanna, it appears to fit the picture quite well. The higher animals are able to live in the savanna the way they do only because the vegetative processes of plant life there are somewhat restrained and subdued in comparison with those of the forest. The animal-plant polarity is weighted more in the former's favor, and not only in the external sense of physical presence. As previously pointed out, highly differentiated mental process goes hand in hand with the inhibition of pure growth activity. Imbuing the savanna with a rich variety of mental process, or "soul life," however, is not the work of the mammals alone, but also of the birds, albeit in a completely different way. The ungulates are the active shapers of the landscape. They act upon their environment by treading it with their limbs and by putting it through a digestive process that entails ingesting it, manuring it, and distributing seeds. It is true that birds are also involved in distributing seeds, their help being essential, for instance, to strangler fig and *Loranthus* species, colorful relations of mistletoe; in addition, sunbirds are important pollinators.

First and foremost, however, birds are the expression in color and sound of the landscape where they live. Frieling (1937) was probably the first to draw attention to this relationship. Drawing attention to it is one thing, but capturing it in words is not so easy, for it cannot be quantitatively "proved," as can the influence of grazers upon vegetation, say, or of predators upon prey. However, since the sphere of activity of the birds' expression lies more in the realm of feeling and atmosphere, it is all the more accessible to *direct experience*. If, therefore, instead of making a sustained effort to observe and take in the

special features of a landscape, we relax our wakeful attention and sur-
render to apprehending the atmosphere by listening for what inner
echoes arise in response to the features, then it will become apparent
that these atmospheric impressions, in a manner beyond the reach of
words, find expression in the musicality of birds, in their calls and
songs. If waking consciousness is allowed to settle in this way to the
level of dreaming sensation then it reverberates inwardly with what
comes to vocal expression in birdsong. Thus, for instance, the high,
keening cries of the Abyssinian nightjar exactly capture the mixture of
unease and melancholy that can overtake one on the edge of an upland
forest at twilight when the cool evening mists roll in, the contours of
the trees become blurred and mobile, and elephants step silently out
of the undergrowth. The exhilarating feeling of release one gets from
finally standing beside a glittering lake thronged with flamingos after
a long struggle through the thorny bush to reach it could not be better expressed than by the joyful triplets of one of the lake's own inhabitants, the African fish eagle.

The point already made about the mutual antagonism between
over-luxuriant plant growth and higher animal life applies also to the
world of birds, as is shown by the *lakes of the East African savanna*.
Lake Naivasha is the only freshwater lake among them and is fringed
by thick groves of papyrus that jut out into it at many points. The
bays are covered near the bank with a thick carpet of water fern
(*Salvinia*), which further out becomes a dense mosaic of water lilies,
each of their large, bright pink blossoms on its stalk a little above the
water. The lake is famous for the richness of its bird life, but it is primarily due to the great diversity of species rather than to large numbers of individuals that it owes its fame among ornithologists. Apart
from the crested coot (*Fulica cristata*), which occurs in large flocks on
the open water, most of the birds are shy and solitary, and have to be
stalked with due caution using a boat. Not a few are also so well cam-
ouflaged and practiced in keeping still that the eye easily passes them
by—the large bulk of the Goliath heron just as easily as the slender
purple heron. Even the pygmy kingfisher with its shimmering blue-
violet feathers merges completely, through its smallness and stillness,

with the confusion of papyrus stalks, and is seen only as an iridescent blue streak as it darts from cover.

At the salt lakes, for example, Manyara, Elmenteita, and above all Nakuru, the picture that greets the visitor is completely different. Except where streams flow in, all vegetation ceases a long way from the shore, giving way to a barren, salt-encrusted beach. The barrenness of the scene is dramatically intensified by the pale skeletons of dead trees, victims of a rise in water level. But the bird life, by the same token, achieves an abundant vitality that is unmatched over the whole Earth. In many places lesser flamingos stand in such dense flocks that a solid pink cover appears to be spread over the lake; their numbers may sometimes touch the million mark (Brown 1959). But quite apart from flamingos, there are such vast numbers of so many different species that one does not know where to look. There is a sense of everything being open, unconfined; radiant white is the dominant color, or, in the case of the flamingos, pink or scarlet highlighted with white. On the shore thousands of pelicans rest, while others paddle on the lake in orderly flotillas, dipping their beaks in the water to a single rhythm and steering around the massive jutting heads of the hippos, which continue to wallow in blissful disregard of the bustling bird life all around them. Some of the skeleton trees sport cormorant nests on every available branch. The parent birds constantly come and go, and the air is filled with the shrill screeching of the chicks begging for food. At the shore, ruffs wade and little stints probe the mud, both winter guests from northern Eurasia and the Arctic. A fish eagle picks at a dead flamingo, while nearby a marabou waits patiently for the leftovers. Near the shore European shovelers dabble in the shallows, and a little further out stand painted storks boring into the mud with their long beaks. Little egrets line the shore like white fence posts, their protruding heads poised ready to strike. But not only the water teems with life. Wave upon wave of birds sweep by through the air, white pelicans with black-hemmed wings in majestic flight, flame-colored clouds of flamingos. In the reed beds beside the lake stand large herds of waterbuck, their stately forms reminiscent of the red deer. Slender reedbuck, which seem here to fill the role of the fallow deer, lie concealed, but their presence is betrayed by the black tips of their horns poking above the reed stalks.

This short glimpse of the salt lakes, the pinnacle of bird-life in Africa, will have to suffice, since to do justice to them in all their fullness we would need to carry on for many pages.

Among the ecosystems of Africa the savanna is, of course, the one most strongly subject to *seasonal change*. Toward the forest on one flank and the desert on the other the change becomes less and less distinct, until finally ceasing altogether in the lowland rainforest or the pure desert. The evergreen rainforest knows no synchronized growth cycles. Instead it is common to find trees on which some branches are shedding their leaves, others are budding, others are bare, and still others are arrayed in their full foliage: autumn and spring, winter and summer, all on the same tree. In the desert the lack of growth cycles is due simply to the complete absence of vegetation. In both these extremes of landscape, moreover, the daily variations in precipitation (in the rainforest) or temperature (in the desert) are much more drastic than the annual variations of seasonal climates. In the Sahara the swing of temperature from day to night exceeds the difference between the summer and winter averages.

With its dry and rainy periods, however, the savanna imposes a seasonal rhythm on plant, animal, and humans, which is strong despite considerable local deviations. As we approach the Tropic of Cancer in the north and the Tropic of Capricorn in the south we enter regions of the savanna where there is only one rainy season, which always coincides with the summer solstice. When the Sun moves from north to south it is followed, as it were, by the rains, which reach the equator around the autumn (northern) equinox (short rainy season). When the Sun crosses the equator again on its return from the south, it brings with it the long rainy season. In this way the rainy season oscillates between north and south, with the result that the regions close to the equator enjoy not only two, but also longer-lasting rainy seasons. Toward north and south both the duration and the amount of rainfall gradually diminish until they fade out entirely. (For example, for the northern savanna, going from south to north, the figures are typically as follows: Sudan-Senegal dry savanna—rainfall 400–1200 mm, 7–9 dry months; southern Sahel—rainfall 250–500 mm, 9 dry months; northern Sahel—rainfall 100–250 mm, 10–11 dry months; after Knapp 1973.) Naturally there are large fluctuations. Even in areas with high rainfall there are extremely dry landscapes, for instance on the downwind side of mountains. Furthermore, the rains are notoriously unreliable and can sometimes fail, not only once but several times in a row, as in the Sahel in recent years.

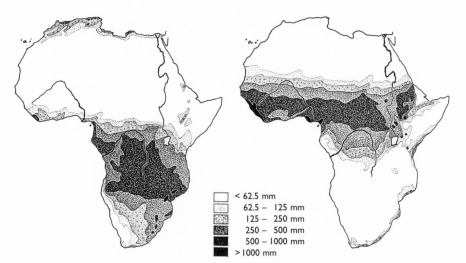

	< 62.5 mm
	62.5 – 125 mm
	125 – 250 mm
	250 – 500 mm
	500 – 1000 mm
	>1000 mm

Fig. 16. Rainfall in southern summer (left; December until February) and in northern summer (right; June until August), total amounts in mm. (After Knapp 1973, redrawn.)

The dry seasons, for their part, appear as countermovements from the periphery to the equator and back, periodic invasions of the center of the continent by the arid climates of the south and the north, particularly the latter. The intermediary position of the savanna between the life pole of the rainforest and the death pole of the desert is thus revealed not as a statically persisting state, but as a process— *the savanna tends rhythmically toward one pole and then the other.*

For the processes of life this rhythm is analogous to breathing. During the drought all life has been "breathed in": the trees have shed their leaves; the grasses all their peripheral organs; numerous amphibians, reptiles, and snails are asleep underground; butterflies are going through their chrysalis phase; the herds have migrated; and certain birds have left for other continents. The air is hazy, dusty, and laced with smoke from the many bush fires that cover the land with charred patches—the black counterimage of our white winter.

These fires are as much an integral part of the savanna as its trees and grasses, herds and pastoral nomads. Until recently it was thought that they were first introduced into the savanna by humans, and that before their appearance natural fires had played a very minor role (e.g., Knapp 1973). Closer investigation, however, yielded a completely different picture. In areas kept clear of artificial fires, it was found that in the drought period natural fires were regularly caused by lightning, and that this did not occur in other areas because the vegetation had already been burned off by human intervention. Moreover, the dry savanna has vegetation that is highly adapted to periodic burning. Thus plants have not only developed ways of protecting themselves from fire but in some cases even depend upon it for setting new growth. This highest level of a "pyrophiliac" plant world has developed most fully in the bushlands of Australia (Göbel 1976; Walter 1973), but the African savanna and the broadly similar cerrado of South America (Eiten 1972) do have fire-resistant vegetation. While fire wreaks irreversible destruction upon the tropical forests, where it cannot occur naturally and is always caused by humans, it does little harm to the savanna. "We must therefore regard fire as an ecological factor that was important in Africa at a time when man did not yet exist. Natural fires have been involved all along in determining the distribution of grassland and woodland. . . . The annual burning of the savanna pasture may be seen as an adaptation of the indigenous population to the fire-climate of Africa" (Walter 1973). Fenced-off areas kept artificially free of fire quickly turned to scrubland (Hendrichs 1971, Knapp 1973). Fire acts upon the landscape in the same way as the wild ungulates and the herds of the nomads. All serve to keep the savanna open and prevent it from turning into an impenetrable, pastureless, sterile tangle of thornbushes, which would provide a habitat for only a very limited number of animals and would entirely exclude humans. Fire is another of the forces that hold back purely vegetative vitality, thus providing the conditions in which the mental life of higher animals and humans can unfold.

Once the rainy season, the "summer," sets in, the landscape visibly "breathes out" in a veritable burst of growth. The winged termites scuttle out of their mounds; it looks like a bubbling spring when the sun glitters on the wings of the dense swarms streaming out of the ground and rising straight into the air, where they form tall columns. Freshwater shrimps are encountered far from their home streams—they, and many other animals besides, are on the hunt for fallen termites. Short-lived puddles are suddenly full of tadpoles, fairy shrimps, and small fish hastily going through their life cycle before the next drought eliminates them (their eggs lie dormant in the ground during the dry season; cf. Burton 1973, Carpenter 1925). Delicate pink fungi shoot out of the ground, the "fruit" of underground termite gardens. But of all the things springing up, it is of course the grasses, the bright pink crinum lilies (*Crinum sp.*), and the large blossoms of numerous other bulbous annuals that take pride of place. Trees and bushes all burst into leaf, and the acacias deck themselves in blossoms that are thronged by bees and beetles. The weaverbirds build new nests and begin to breed. The whole animal world begins to teem with newcomers. Wildebeest calves gambol around for the sheer joy of it while the herds roam through the fresh grass; zebra foals frolic, and gazelle fawns try out their first awkward jumps on their long, matchstick legs. And for the human population the time of sowing and planting has come.

The savanna is also the homeland of the tribes who have formed the closest ties to animals. These *pastoral peoples* make cyclical migrations with their cattle and, like the wild ungulates, follow the seasonal rhythm of rainy and dry periods. Of course, not all savanna dwellers are pastoral, but the cattle breeders are certainly the most characteristic and impressive among them. In ethnic terms they are the most colorful element in the savanna. The full picture is very tellingly described in a myth of the Kikuyu, a Bantu tribe of farmers native to East Africa. It tells how the creator-god, Mumbere, had three sons. He laid three tools before them and bade them choose one. The first, called Masai, chose the spear, the weapon of the warlike shepherds, the second chose the bow, and the third, who became the chief ancestor of the Kikuyu, chose the digging stick (Britton and Ripley 1963).

This story accurately names the three cultural groups of the savanna. The oldest layer is that of the *hunter-gatherers,* whose most important tool is the bow and poisoned arrow. Today only pockets of them survive among the *pastoral nomads* and the *farmers.* Strikingly enough, even today, in spite of much mixing and overlapping, each of the three cultural groups has its own distinctly discernible ethnic characteristics. The hunter-gatherers, surviving only in scattered groups in remote arid regions ranging from South Africa to Ethiopia, will be found to have Khoisanid (Bushman) traits. In many cases the pastoral nomads show Hamitic ancestry, with their long limbs and sharply contoured faces. In a number of tribes also the Negroid element has, through mixing, become very pronounced (Masai). Others have preserved their original features more strongly, so that in many parts of East Africa one encounters people with skin so pale-brown that it looks almost European, for example the Tutsi or the Iraku of northern Tanzania, who in their bearing and ancient language are thought to have preserved proto-Hamitic features (Murdock 1959).

Since the primeval element—the hunter-gatherers—has disappeared from most regions, especially those most densely populated, the savanna belt is largely characterized by the juxtaposition of two fundamentally distinct cultural groups, the pastoral nomads and the farmers. Among the former, Hamitic traits tend to predominate, while the latter have more black African features. In other words, the distinction is not only cultural, but also has a marked ethnic component (by this is meant the predominance of certain racial characteristics; there can be no question of pure races). This applies to northern Nigeria and Cameroon, where the Negritic farmers share the land with the pastoral Fulba, just as much as to the lake region of central Africa (Hamitic Tutsi in conjunction with the farming Bantu tribes of the Hutu) or the East African plateau (pastoral nomads such as Masai and Iraku beside Bantu farmers such as the Jagga or the Kikuyu). Naturally the two groups have influenced each other here and there, creating overlaps and exceptions—Negroid cattle breeders, or Masai farmers such as the Arusha in northern Tanzania. In general, however, the rule holds good (cf. Hirschberg 1965).

In tune with this pattern it is also perfectly natural that the herding peoples prefer the drier parts of the savanna where grassland predominates, while the farmers choose to live in regions with higher rainfall. This means that the black farming peoples have tended to settle in areas where the savanna adjoins the forest, into which they have gradually made inroads, whereas the herding peoples have favored the land that at its extremities merges into the desert. In the mosaic that constitutes the landscape of East Africa, this cultural divide is very striking: the arid grasslands of the Rift Valley are where the Masai graze their cattle; the upland forest regions are home to the Kikuyu, Jagga, Pare, Teita, etc. The thorny wilderness of the driest bush and the semi-desert is shunned by both groups and has thus become the last refuge of the Khoisanid hunters, for example the Hazapi of Lake Eyasi.

The *black peoples* are skilled *husbandmen,* and their style of cultivation in no way undoes the forest character of the landscape, but rather transforms it. The stands of banana and papaya in this "cultivated forest" are interspersed with single tall forest trees, which break the otherwise uniform appearance of the landscape. Particularly impressive is what meets the eye in the monsoon region close to the East African coast. There the intensely cultivated land of the Giriama looks like a well-tended primeval forest, or else a very lush, dense park, a setting in which the unmistakable forms of individual trees stand out to best advantage. It is a delight to travel through such a landscape and see how caring human hands have shaped something diverse and vibrant with not a trace of the monotony typical of the natural forest, the bush, or many other parts of the savanna. Tall thin coconut palms alternate with the rounded mountainous masses of mango trees. In between them stand the geometrical cotton trees with their straight trunks, each with its helix of right-angled branches, and the round tops of the cashew trees glowing with red, pear-shaped fruits with their funny appendages, the curved nuts. The greatest contrast to the fragile-looking, but nonetheless supple palms is provided by the gigantic barrels of the baobabs, which occupy the gaps in the vegetation. The farmers' whitewashed houses with their roofs of palm leaves are scarcely visible among the filigree of foliage, branches, and tree trunks.

In the regions where the *pastoral nomads* roam with their herds, however, there are scarcely any trees. The Masai Steppe is to a large extent pure grassland, and there is a growing danger that the pasture will be completely overgrazed by the ever-increasing herds, thereby giving the desert a chance to take hold. Indeed the nomads of the savanna have in their nature features already familiar to us in the desert nomads. Among them the tribal customs are still so much alive and mold each member of the group so strongly that it is justifiable to speak of a common "tribe mentality." This colors the behavior of each *Masai* to a greater extent than any personal traits. Among these tribal features is their tendency to regard themselves as an elite, an austere attitude that places them in cultural isolation and makes them aggressive toward all other tribes. The purest expression of this is the *moran,* or warrior, who leads a free-roaming existence with no social obligations apart from defending the group and rustling cattle. This basic attitude of self-sufficient isolation operates also in relation to the natural environment, from which the Masai take almost nothing, not even game. Their whole life is built around cattle herding and they have just as close bonds with their animals as do the desert nomads.

If the world of the Masai and other pastoral nomads is strongly masculine, shaped by the qualities of the warrior, the culture of the black farmers, by contrast, is much more feminine in its basic features. This is evident not only in their careful tending of crops, but also in their basic nature, which is softer, more sensitive and accommodating to outside impressions and influences. Their basic turn of mind could be characterized as "sympathetic," in contrast to the "antipathy" that sets the tone among the Hamitic pastoral tribes (Suchantke 1972).

The sharp, angular features of a Hamitic herdsman appear old and well seasoned in comparison with the soft lines of a Bantu face. Their emaciated forms, stoical stillness, and emotional coolness are physiognomic and psychological expressions of old age, as is their lack of attachment to nature. They are the conservative preservers of their old traditions, whereas the Bantu are the youthful protagonists of the turbulent development gradually taking hold of Africa.

Thus it is clearly the case that the two poles between which the savanna mediates, namely the desert and the forest, not only figure in the face of the landscape, but are also reflected in the typical features of its human inhabitants. In the savanna human life has been able to unfold more richly and diversely than anywhere else in Africa. In this connection the African savanna occupies a uniquely significant position in the world—it is in all likelihood humankind's place of origin.

Historically, scientific opinion first centered upon South Asia as the point of origin (because of the *Pithecanthropus* finds in Java), but it was not long before evidence began to build up in favor of the idea— put forward as early as 1910 by Rudolf Steiner—that Africa was the prime location. Paleontologists quickly came to regard the *Australopithecus* specimens uncovered in South and East Africa as the direct ancestors of humankind. Subsequently, however, they were classified as an evolutionary side branch (Gieseler 1974; Heberer 1974). With their strange mixture of apelike and human features, they were considered not prehominids, but protohominids (Portmann). They were apelike chiefly in the form of the skull, and humanlike in the structure of pelvis and legs that enabled (probably not habitual) bipedalism. More important were the later finds of skeletal remains providing evidence that genuine hominids (*Homo habilis*) existed at the same time. Recently at sites by Lake Rudolf and in Ethiopia there have been further extensive finds that prove the existence of fully bipedal contemporaries of *Australopithecus* reaching beyond the Ice Age right back into the Tertiary period (R. E. Leakey 1973).

III. SUMMARY. HUMANKIND AND THE FUTURE OF THE AFRICAN LANDSCAPE

In conclusion let us review our impressions. The African continent can best be understood as a large dynamically polarized threefold organism. Between the life pole of the rainforest and the death pole of the desert, the *savanna* acts as mediator, combining and balancing the two extremes. In a cycle of rhythmic processes the harshness of the dry season is softened by the burgeoning of life that follows the rain, while in turn the explosive growth of the rainy season is subdued and driven back by the onset of drought.

The savanna also occupies a middle position geographically, mediating between the desert, which dominates the northern, northeastern, and southern margins of the continent, and the rainforest at the very center. Only in the east do the three landscapes intermingle in a rich mosaic in which both desert or semidesert areas and moist upland forests are surrounded by wide expanses of savanna of all types, from dry forest to pure grassland.

The savanna is in a certain sense the most African of the three landscapes. The rainforest is not only the most weakly represented of the three in terms of both area and the relative diversity of its life-forms, but it also has the strange feature of being restricted to the west and west center of the continent, in fact to the area around the large bay where what is now South America once lay when the two continents were still joined. Although the flora and fauna of the African rainforest are largely specific to it or are more closely related to those of other continents (Thorne 1973), it nevertheless appears as an impoverished, eastern offshoot of the Earth's chief rainforest in South America.

The opposite applies to the Sahara, which clearly has links to the east and carries on into the Middle Eastern and South Asian deserts as far as India. Many of its animal and plant species enjoy a similarly wide distribution (for birds: Moreau 1966; for mammals: Heim de Balzac 1936).

The three landscapes can also be understood as three phases in the aging of the continental organism. The rainforest is dominated by processes typical of the juvenile growth phase of an organism. In this environment, therefore, organisms that have progressed to the development of conscious sensory awareness—a faculty that always goes hand in hand with the inhibition of living processes—are confronted with conditions entirely unfavorable to them in that they are exposed to influences that would keep them at the juvenile level.

In the savanna this vegetative tug of the rainforest is removed and we see the African landscape at the peak of maturity, with pure growth processes held within bounds and balanced by the presence of diverse creatures endowed with a wide range of mental capacities. The landscape itself goes through processes akin to the breathing in and out, sleeping and waking of its animal inhabitants: the regenerative burst of plant growth and animal reproduction in the rainy season is followed by the dry season in which vegetation dies back and reproduction is dormant.

The closer we get to the desert the "older" the landscape becomes. The seasonal rhythm slows down and finally peters out, breathing stops. In the absolute desert all plant life has died out. Animals, the mind of the landscape, have also departed, or appear only as occasional visitors able to pit their self-sufficiency against its rigors. The flowers of human culture that grew up here have long ago been uprooted and transplanted to other, later cultural milieus.

Unfortunately it cannot be said that these three "age phases" of Africa stand in a balanced relationship to each other. It would seem, rather, that the continent as a whole is becoming older. This expresses itself in the advance of the desert and the retreat of the rainforest. There is evidence that this has been occurring naturally since the end of the Ice Age, but the process is now being greatly accelerated by human activities. To conclude, therefore, we will take a closer look at this aging process, since it is of crucial significance for the future of Africa.

Forest destruction is most extensive in West Africa, where shifting agriculture has been practiced for a long time by indigenous peoples, who have gradually "wandered" through vast tracts of land. Intensive modern agriculture has now been added to the picture, together with a form of commercial forestry that pays scant attention to local ecological conditions. This modern tropical forestry proceeds either by altering the composition of the forests through applying herbicides to commercially worthless species, or by felling the upland forests and replacing them with monocultures of fast-growing, non-African conifers (pine, cypress, etc.). In Africa things are no different than in other continents: no landscape, no ecosystem in the world is so imminently threatened with complete destruction as the tropical forest. The consequences for Africa (as for other continents) are clearly discernible: increasing aridity. Events are taking their inevitable course, even though as early as 1949 there were warnings that the rainy season would shorten and the amount of rainfall lessen with the shrinking of the rainforest, since it constantly feeds the atmosphere with vast quantities of moisture: "We may compare the climatic effects of the rainforests with that of the ocean. They determine the level of rainfall and thus play an important part in the water economy of the continent" (Aubréville 1949).

Against this background the immediate future of the savanna, and with it the future of Africa, looks anything but hopeful. The similarity between the savanna and the agricultural landscape of Europe is, as we noted previously, not merely skin-deep. Historically this "cultivated savanna" also comes between forest and desert. It was originally wrested from the forest by humans. In the Mediterranean, where this cultural process began much earlier than in central and western Europe, we can see what overcropping made of this landscape: an eroded wasteland. But the process need not necessarily follow this course; the desert begins to set in only when the forest has been eliminated from the landscape instead of being integrated into it in an appropriate form. The forest stores moisture and is thus a landscape's chief source of regeneration. If left to itself—and this goes for the forests of the north as much as for those of the tropics—it will outgrow and displace everything else, whereas if it is altogether absent all life will drain out of the landscape.

Perhaps the most important quality the savanna and the cultivated landscape (in its ecologically sound, properly husbanded state) have in common is that both are landscapes in which plants, animals, and humans can live together in an optimal way. In the latter case this is the outcome of human action. And it was through human action based not upon the ruthless exploitation of our fellow creatures but upon a nurturing, caring attitude that an evolutionary enhancement of nature was achieved. Perhaps the method involved was "prescientific," but on that account it was all the more sound in its instincts. No doubt certain species were displaced or even eliminated, but others received new habitats and became associates of human culture: many birds from the stork to the starling, numerous mammals from the deer to the hedgehog, as well as innumerable lower organisms. In the human-made meadows and fields, new species of plants emerged spontaneously through crossing with wild plants that invaded the new landscapes from distant regions (Landolt 1970).

But in Africa, the land of the natural savanna—and in other tropical continents—what is to be done? Is there really only a choice between destructive exploitation and conservation in the form of inviolable nature reserves? This would not offer a very hopeful prospect, for the creation of reserves is no lasting solution, and ultimately is nothing other than a negative reaction to the destruction of nature. It would be an admission of bankruptcy on the part of our civilization if it had no other options in its relations with nature than either total destruction or complete renunciation of any kind of interference—quite apart from the fact that such sanctuaries, as is already well known, do not long remain intact, but need care and attention if they are not to degenerate.

This should not be misunderstood as a plea against the creation of nature reserves. In the short term they are the best, perhaps the only, way of saving certain life-forms and habitats, creating opportunities to study them, and awakening love for nature in large numbers of people. As such they deserve our wholehearted approval. But it would be an illusion to think that they offered a long-lasting solution.

That can only be achieved in another way: through a thoroughgoing partnership between humans and nature in which nature could be developed as well as protected and cared for. The ethos of modern scientific, political, and economic life is largely incompatible with such an idea; the dominating powers, to whom we owe our environmental ills, are the most unsuited and the least inclined not only to repair the damage, but also—and this is the crucial point—to develop alternatives.

For the promise of success we need to look in quite different directions. One is the relatively new science of ecology, which studies living systems not by analyzing them into their constituent parts and thereby losing sight of the connecting thread, but by focusing on the relationships between the parts as evidence of a higher level of organization. Another is to be found in various methods of traditional agriculture. It is striking how much respect ecologists have developed in recent years for indigenous methods of cultivation that used to be contemptuously dismissed as primitive. Now ecologist have recognized that science has something to learn from such intuitive practices.

The farmers of the Usambara Mountains in Tanzania, for instance, have developed a system of mixed cultivation that makes optimum use of the nutrient-deficient soil (Egger 1975; Egger and Gläser 1975). Tall shade trees and banana palms are mixed in among smaller plants, all in a chaotic profusion. To the Western agricultural specialist used to the intensive application of chemicals and technology to crops growing in ordered rows, the picture this presents is that of a neglected wilderness. It is this very diversity, however, and the healthy naturalness of this style of cultivation that impress the ecologist. And a person with even the merest touch of aesthetic sensibility will be attracted by the beauty and variety of the landscape. "We met the owner and inspected the field with him," writes K. Egger (1975). "Why, we asked him, do you leave the weeds to go to seed? Grinning at such a stupid question, the farmer explains, 'Otherwise there wouldn't be enough new growth!' What do you need them for? In reply to this we were treated to a lecture in ecology. 'The weeds shield the soil, whether they are growing or newly mown. That's the reason there is no erosion here. The sun makes the soil hot; with the weeds it has shade and doesn't dry out. The layer of mown weeds keeps the soil soft and moist right up to the surface. It rots slowly and feeds the young seedlings. It's enough if I give my potatoes a bit of cow manure; that's plenty of fertilizer for me!'"

Here there is no need of artificial fertilizer or insecticide. What appears to the "expert" accustomed to monoculture as an untended chaos turns out to be a healthy, skillfully husbanded living organism, "a stratified form of cultivation perfectly patterned upon the structure of the primeval forest that preceded it" (Egger). Monocultures stand in sharp contrast to this, especially in the tropics, where they are highly vulnerable to pests and drain soil fertility very rapidly. The massive application of artificial fertilizer they require is much too dear, and anyway most of it ends up being washed away by the rain (Egger 1975; Sioli 1973). What is more, monocultures tend to promote formerly benign animals and plants to the status of pests. This pest breeding is then intensified by the application of costly pesticides, which select for resistant strains. Just who benefits from these environmentally lethal measures is all too obvious.

The unprecedented concentrations of economic and political power that stand behind modern intensive agriculture and the loss of tradition and uprooting of the population associated with it are fast doing away with the old, time-honored methods. But these should and could be adopted and further developed, and there is even a tiny spark of hope for this option if the change in attitude of recent years away from exploitation of nature and toward cooperation becomes sufficiently widespread. A fruitful partnership could emerge, combining refined and time-tested indigenous methods with modern ecological know-how and new agricultural methods closely akin to it, such as the approach known as biodynamic agriculture. Such an alliance would, of course, be completely at odds with the way things have been done up to now. Instead of simply transferring the methods of the industrialized world to "underdeveloped" regions, it would mean taking local knowledge seriously and working together with those in possession of it, rather than looking down on them. This approach would open up totally new possibilities.

The finely tuned expertise of the East African forest farmers described by Egger could form the basis, upon which others could build, for a way of cultivating tropical rainforests while preserving their specific ecological structure. There is an extreme urgency about this, for something like it may be the only way of escaping from the current destruction/conservation pattern, which is not a sustainable alternative. Through the integrated approach of ecologically appropriate husbandry, the tropical forests could be used in being preserved, preserved in being used, and this would be of truly worldwide significance. It would mean that the progressive drying out of Africa, and indeed of other tropical continents, would be counteracted. But there could be still other consequences: taking indigenous cultures seriously could lead to the discovery and development of new domesticated plants, of which there are many in the early stages of domestication in the fields of Africa (Knapp 1973) and South America (Brücher 1968). But with the loss of tradition on the part of their breeders, they are in danger of disappearing.

Although this would prevent the African savanna from drying out further and drifting toward the desert state, it would not provide the conditions for its gradual transformation into a healthy cultivated landscape, at least not one that could accommodate the abundant animal populations with their highly refined adaptations to climate and vegetation. In this regard the existing local cultures, be they those of nomadic cattle breeders or of farmers, offer no useful leads. The fact is that Black Africa has not produced one single domesticated animal, which is astonishing when one considers the sheer abundance of large mammals (and birds) this continent contains. Clearly the pastoral nomads have shown little creativity, at least in this matter, and the black farmers would seem to have a much stronger relationship to plants and their cultivation than to animals and stockbreeding.

This may well be the reason why the oft-repeated suggestions on careful cropping of the wild populations of the savanna (e.g., Huxley 1965) have so far not been taken up in black Africa, and have been tried only in white South Africa. Experience shows that not only large species like eland and zebra but also smaller ones like impala are well suited to open game ranching (Bigalke 1974). Wildlife management of this sort would be ideal for all those vast regions of the African savanna where, due to dryness, cattle breeding is unsuitable and any attempt at arable cultivation only furthers desertification. Moreover, experience once again shows that it would be economically viable (Vincent

1974). That international conservation thinking is tending in this direction is a good sign, and it is to be hoped that such proposals will find open ears in the relevant African government departments.

These examples show that Africa is not irreversibly heading toward a bleak fate, that the continent is not an aged organism inexorably succumbing to death, in other words, desert formation. They show, rather, that it is up to humankind what future Africa will have. If human beings could become a force for rejuvenation and renewal in Africa, working with the potential already supplied by nature in anticipation of just such further development by human action, then the future of Africa could be bright indeed.

What Do Rainforests Have to Do with Us?

GLOBAL THINKING—GLOBAL RESPONSIBILITY

The idea of the nature reserve originated in North America. It may be that its first stirrings were present in other countries at about the same time, but it was the United States that took the first steps toward putting it into practice with the founding of National Parks—Yellowstone in 1872 and Yosemite a little later.

This action was prompted by the desire to preserve something of the unspoiled natural beauty of America—not for its own sake, but for the benefit of human beings. Underlying this was the feeling that human beings needed not only physical nourishment for the body, but also *aesthetic* nourishment for the senses, and that their sensibilities would be enriched and enlivened by direct contact with grand and varied natural scenery. This is what is superficially referred to as the "recreational value" of a landscape. Nevertheless it implies recognition of the fact that a landscape's aesthetic value is just as important as its material yield of minerals or foodstuffs.

In the meantime things have changed so much that there have long been more serious reasons for conservation. We now face the question of whether life on Earth can continue to exist at all if we do not begin treating nature with more care and respect. What is more, regional or national thinking is no longer adequate, for the problems and crises have taken on worldwide proportions and have little chance of being solved without the concerted action of humanity as a whole. How else could the greenhouse effect, the hole in the ozone layer, the pollution of the oceans, and the destruction of both tropical and temperate forests be combated? Precisely the global scale of these problems renders any kind of national or narrowly individualistic approach to them, in thought or deed, obsolete.

Humanity—in its own interest—will have to get used once again to a thriftier, more sustainable lifestyle. The fact is that ultimately the needs and interests of humanity cannot be separated from those of nature. What is good for one is good for the other. It is time to remove the artificial barrier we (in the West) have erected between culture and nature. Other cultures (including past phases of our own), commonly dismissed as "primitive," provide instructive models of how nature can be cultivated in ways that are both ecologically sound and economically successful. In what follows this will be shown to apply even to the rainforest—of all ecosystems the one widely held to be so sensitive that any kind of interference threatens it with irreparable damage.

The idea that the rainforest is of purely local significance—it covers after all only 3% of the Earth's land surface—is also wrongheaded. It proves untenable when attention is focused upon the connections and mutual interactions among the Earth's living systems. The fact is that those 3% account for almost one-third (29%) of terrestrial plant biomass, which represents a truly vast amount of fixed carbon. This is now in the process of being "released" into the atmosphere. Forests, especially those covering large regions with an unbroken canopy, exert a large-scale influence upon climate (Prance 1986). They store enormous quantities of water, and release it in small doses, through transpiration, to the environment at large. Thus they prevent extremes of temperature and keep the air moist, just like a large body of water. The lack of this regulatory activity is already being sharply felt in southern Brazil, where only about 1% of the original forest cover of the state of Sao Paulo is still standing. For instance, in 1975 the greater part of the coffee crop succumbed to frost as a result of a front containing Antarctic air that pushed northward into the region. Many people were thrown out of work, and coffee prices rose. (This wave of cold actually penetrated a lot further north—I was in a party visiting the Upper Amazon at that time, and night temperatures of 12°C caused us considerable discomfort.)

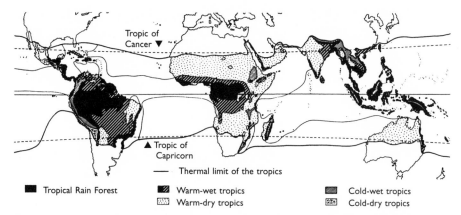

Fig. 1. Hygrothermic distribution and configuration of the tropics. Warm-wet: original extent of the rainforests (black); moist savanna (broad diagonal hatching). Warm-dry: deserts and semideserts; thorn forests; dry savannas. Cold-wet: mountain and cloud forests. Cold-dry: mountain deserts (for example in the Andes). (After W. Lauer 1989.)

The rainforests of the Congo and Amazon basins resemble gigantic saturated sponges. In their root systems they retain a good part of the daily downpour, and they absorb the floodwater that streams out of the mountains in the rainy season or when the snow melts. The Amazon catchment area contains 18% of all the freshwater that flows into the Earth's oceans, in other words, almost one-fifth. It is not difficult to imagine what will happen once the forests have finally disappeared: floods and erosion will make vast regions uninhabitable, turning them eventually into desert.

Of course, such events might not have to wait so long, and indeed have already occurred in not a few areas. Astonishingly enough, local populations are likely to be unaware of the roots of such disasters. This may be simply because of ignorance, but unfortunately indifference and irresponsibility cannot be ruled out, for although local cultures can usually be expected to have developed over generations the ability to live in harmony with nature, this is not necessarily the case where they have been subject to colonial or modern industrial influences.

The Indian farmers of the Andean plateaus certainly provide a splendid example of harmonious development. Their agriculture is very well attuned to local ecological conditions (Franquemont 1990;

Gade 1975). In the South American rainforest, however, there have been no indigenous agricultural communities, apart from a few notable exceptions, to which we will return later. Moreover, the indigenous peoples are facing extinction at the hand of outside pressures, and the people settling in their place are "Neo-Europeans" (Crosby 1988). They bring with them exploitative forms of agriculture totally foreign to the natural landscape, such as the plantation, which in the past was tended by slaves but now depends on migrant workers and mechanization. These Neo-Europeans have so far failed to establish any true connection to nature in their new homeland. Its ecology remains a closed book to them.

This local ignorance stands in stark contrast to the fact that the ecology of the rainforest is now very well known. So well known, indeed, as to provide a basis for its sound and sensitive management, which could even include cultivating it in ways that would be beneficial for both nature and humans. But before we turn to this point it is important to form some idea of what the rainforest is, of its intrinsic nature as a composite whole. To make this possible we must begin by looking at some of its special features.

MONOTONY IN GREEN

Upon first approaching the edge of the rainforest, for instance by river, it appears completely impenetrable, all ways barred by bushes, trees, and giant ferns, matted together by lianas and a surfeit of epiphytes clinging to every branch and trunk. Behind this tangled facade the forest interior appears enveloped in profound darkness. Of course, what the eye here observes is not the actual forest, but the structure by which it seals itself off from the outside world. Temperate forests present a very similar picture where their margins have been left undisturbed: a wall of green with never a chink from the ground to the treetops. A forest has its own identity as an organism, and while it interacts with its environment, like any other living thing, at the same time it separates itself off by means of a boundary layer or "skin."

Fig. 2. Forest edge where it meets a stream in the Serra do Mar of southeastern Brazil in 1977—a memento, as this forest has long since been destroyed. Epiphytic bromeliads can be seen on the left (above and below) as well as arums (above right) or their hanging air roots, up which a climbing fern *Lygodium* is growing. Other members of the arum family are winding their way up the trunk of the tree fern. (From Suchantke 1982.)

Thus going into the forest interior is much easier than its "skin" would lead one to expect. Apart from the twists and coils of lianas here and there and the hanging aerial roots of epiphytes, there are no major obstacles. Undergrowth makes a rather feeble showing because of the dimness of the light. It has been worked out that in some places only one one-thousandth of the light falling on the forest canopy reaches the ground (Bünning 1947). Even the densest temperate deciduous forest is not nearly so dark. Massive smooth trunks strive upward to the light, branching out only when they have reached a great height. Those that fail to make it remain thin and stunted, topple over, and die. Other plant forms attain the heights either by climbing up tree trunks, like certain species of arum, ferns, and even cacti, or by occupying the light-filled canopy from the very outset, like the innumerable epiphytes—especially bromeliads, arum species, ferns, orchids, and of course lichens and mosses. There are even trees that germinate in the airy upper regions and use other trees as props: the strangler figs. Their seeds are deposited by birds in forks of the upper branches; after germination they send aerial roots down along the trunk of the host tree, which branch, intermesh, and gradually form a solid coat of armor around the unfortunate prop. The host tree is prevented from growing any thicker and eventually dies, leaving the strangler fig, which in the meantime has rooted in the soil and sprouted its full complement of branches, standing in the original tree's place.

The world of the forest is monotonous. Hours of walking produce scarcely any change of scene. This makes it all the easier to lose one's way, for in spite of the loose density of trees, the field of vision remains very restricted, especially as the forest canopy permits only fleeting glimpses of the sky. There have been cases of people who entered the forest intending only to take a short stroll and never returned. There are no encounters with large animals, and while constant cries from the treetops betray the presence of birds and monkeys, none of them ever becomes visible. Finding a blossom provides very welcome relief for senses saturated by the endless monochrome of the lush foliage.

The persistent notion of the rainforest as a paradisial riot of blossoms was challenged long ago by Alfred Russell Wallace, the contemporary of Darwin and famous explorer of the Indonesian islands and Amazonia.

> In vain did I gaze over these vast walls of verdure, among the pendant creepers and bushy shrubs. . . . [N]ot one single spot of bright colour could be seen; not one single tree or bush or creeper bore a flower sufficiently conspicuous to form an object in the landscape. . . . Many [tropical flowering plants] are very rare, others extremely local, while a considerable number inhabit the more arid regions . . .

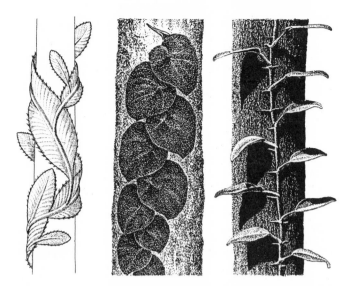

Fig. 3. Tree-trunk–climbing plants of the South American rainforest. At left the spiraling cactus *Strophocactus wittii* (from Rauh 1979); the others are representatives of the arum family. Under its clinging leaves the plant at center collects humus, which it then penetrates with its roots (from Suchantke 1982).

in which tropical vegetation does not exhibit itself in its usual luxuriance. Fine and varied foliage, rather than gay flowers, is more characteristic of those parts where tropical vegetation attains its highest development. . . . The result of [my] examinations has convinced me that the bright colours of flowers have a much greater influence on the general aspect of nature in temperate than in tropical climates. During twelve years spent amid the grandest tropical vegetation, I have seen nothing comparable to the effect produced on our [English] landscapes by gorse, broom, heather, wild hyacinths, hawthorn, purple orchises, and buttercups. (Wallace 1869)

Blossoms abound only in the upper layer of the forest canopy where the rich epiphyte vegetation roots, e.g., full-bloomed orchids. Here there is enough light to create the right conditions for flowers and the birds and insects to pollinate them.

Many rainforest blossoms reveal upon closer examination that the brightly colored (usually bright scarlet) parts that stand out are not petals at all, but sepals or bracts lying below the actual blossoms, which are small and inconspicuous. The blossom itself cannot attain its full expression; it is drawn down into the vegetative leaves and "swallowed." It would seem that blossoms cannot compete with the vegetative vitality of the lower leaves: many trees only blossom once they have shed their leaves!

MORPHOLOGICAL CONVERGENCE

To come into the forest expecting to find a wide range of different leaf forms will also lead to disappointment, for here again monotony is the order of the day. Most trees have almost identical leaves, and it is impossible to use them as a basis for distinguishing between species. On first view such uniformity gives no hint of the tremendous species diversity underlying it: every ten hectares contains up to four hundred tree species, belonging to a wide range of different families (Bourgeron 1983) (Fig. 4, below). In this respect the rainforest contrasts sharply with the deciduous forests of temperate regions. Even in wild mixed forests of central Europe no more than ten species of tree can be expected, but most of these will differ markedly in terms of growth form and leaf shape: oaks, beeches, elms, maples, etc. Major similarities occur only in closely related species, for example pedunculate oak and sessile oak, summer lime and winter lime. In the rainforest, however, what we have is the uniformity of the unrelated. *Environmental pressures dictating uniformity of structure* seem to be stronger than genetic differences peculiar to individual species.

The same picture emerges from a survey of the butterflies and birds of the rainforest. Butterflies are abundant; they prefer the humid shade of the forest interior (and margins) to the baking heat of the open landscape. But in spite of the great variety of species considerable convergence in color and markings has taken place. There are differences, but they correspond to different levels of the forest, such that all butterflies belonging to a particular level wear the same "uniform" (Fig. 4).

The tropical forest displays a characteristic vertical structure in being divided into a series of layers. There is one at ground level, and

Fig. 4. Isomorphism (convergence) of species that are not closely related in butterflies and plants of the rainforests. Below: Leaves of representatives of 26 different species of plants (from Vareschi 1980). Above: typical representatives of two communities of butterflies occupying habitats one above the other: the transparent complex (below) and the rust-brown "tiger" complex (above). Both complexes are very rich in species; only a small selection is shown, including examples of four different families of each complex. Included in the tiger complex are above left a swallowtail, below it a danaid, below right a species of *Heliconius*; the others are ithomiids. The transparent complex includes, besides the ithomiids that dominate here also, two whites (the two specimens in the middle of the bottom row), a nymphalid (above right), and a day-flying moth (below right). (See also Fig. 5.)

in the canopy there are lower, middle, and upper layers. In each layer the light and color conditions, and the associated vegetation, are different. The higher the layer the more light it will receive and the richer it will be in flowering, herbaceous epiphytes. The layers here arranged *one above the other* also occur in temperate forests, but there they are found *side by side*. In terms of this correspondence, then, moving upward in the tropical forest is equivalent to moving out from under the trees into the flower meadow. And just as in temperate latitudes there is a color contrast between meadow and forest-loving butterflies, so here we find the same phenomenon, only one layer is above the other, instead of side by side. The contrast is also much stronger in the tropics, while only a trained eye can distinguish the various species occupying the same stratum.

The butterflies of a particular layer, however, not only resemble each other, but to an amazing degree, as exact measurements have shown, they also match the colors and the intensity, as well as the relative distribution, of light and shade in their surroundings (Papageorgis 1975). The same goes for the birds, in which we find the dull, dowdy ones closest to the ground, mainly light reddish-brown forms at the halfway level, and in the treetops the strikingly colored species, whose dappled plumage renders them invisible in the flickering light and shade of the sunlit canopy (Koepcke 1973) (Fig. 5).

The *formative influence of the environment* overrides the expression of traits typical of individual species. Intolerant of any kind of separateness, the rainforest subsumes everything into itself, and so it is not amenable, for instance, to warm-blooded animals with their natural tendency to emancipate themselves from the environment through the development of a complex "mental" life. Thus the animals that here come into their own, in an incredible explosion of diversity, are the insects. They are not noted for their rich mental life, but instead are perfectly geared to their environment through powerful instincts, which act in complicated, but rigidly determined, almost mechanical ways. Insect "camouflage," by which they, as it were, "become" their environment, is the living image of this. It appears in the play of color, light, and shade reflected in the wings of butterflies; in the way

Fig. 5. Diagram of the vertically stratified winged populations of the rainforest.

Left. Butterfly species of similar coloration: A) transparent; B) rust-red; and black "tiger"; C) black-red; D) blueblack-white complex. (Graph: horizontal axis indicates altitude in meters; vertical axis indicates population density). It is clear that these groups exclude each other in spite of some overlapping.

Right: Similar plumage coloration in birds inhabiting the same altitudes. Below: blackish-dark ground birds: 1 tapaculo; 2 thrush. Above them: rust-brown species with dark speckles, of similar coloration and from the same altitude as the "tiger" complex of butterflies: 3 hummingbird; 4 tyrant; 5 wood-creeper. In the upper canopy glaring contrasts predominate: steel-blue and gold-yellow, or black and coppery gold in the tanagers, 6 and 7; emerald-green with a scarlet-red head and wing coverts in the parakeet, 8. (The birds pictured here are only examples of species-rich groups). (Combined from Koepcke 1973, Papageorgis 1975, Suchantke 1982.)

grasshoppers mimic leaves, not only with veins and ribs but even with chunks eaten out of them and fungal damage; in moths that look just like wilted leaves; in stick insects that imitate dry twigs; etc.

Organisms emancipated from their environment, to whatever degree, will have a hard time in this one. This applies above all to animals with refined mental capacities, which means chiefly the mammals, but also includes human beings. The main characteristic of mammal evolution is the gradual development of an increasingly rich "second nature," which is highly differentiated and not so much related to the immediate natural environment as it is involved in the world of relationships set up within the life of a group (wolf pack, troop of monkeys, family of elephants, etc.). It is no surprise, then, that highly evolved mentally endowed animals are the exception rather than the rule in the rainforest. And, moreover, many of those that do occur tend toward dwarfish proportions (the hare-sized pygmy antelope— the smallest in the world—does not run with its relations in the savanna, but keeps to the forest). This is particularly true of Africa, where the savanna abounds (or abounded) in large animals. In the few genera with species in both forest and savanna, the forest-dwelling species are always smaller: for example, forest buffalo, forest elephant, pygmy hippopotamus. The same goes for many forest-dwelling peoples, the most extreme example among these being the Pygmies. The interesting thing here is that the forest-dwellers are not simply diminutive versions of their brothers in the savanna, but retain youthful proportions even as adults. This feature is also detectable in forest animals, but it is much more striking in humans than in buffaloes and elephants. In the course of their development the Pygmies do not go through the second growth spurt that changes bodily proportions from those of a child to those of a (non-Pygmy) adult.

Food scarcity is also part of the picture here, for both animal and human being. Large ground-dwelling animals are at a disadvantage because the richest source of food is in the treetops (which is where most forest animals live). It has been observed that numerous rainforest mammals, as diverse in form as kinkajous, bats, armadillos and sloths, have a lower metabolic rate than animals of other landscapes. Some species, such as anteaters, armadillos, and sloths, "do not reach . . . even half the normal metabolic rate of mammals in their size-range. In the giant armadillo the rate is down at a level (29%) closer to that for reptiles than for mammals" (Reichholf 1990). In contrast to the grasslands with their abundance of cereals, the rainforest is especially poor in protein-rich food, and since the animal body is built up largely out of protein this, apart from any other considerations, places immediate constraints upon body size (and population size—there is nothing remotely comparable to the herds of the open landscape).

Understandably enough, the rainforest also sets limits to the unfolding of human culture. The extremely harsh living conditions, the shortage of food, and the impossibility of developing any kind of viable agriculture prevent populations from becoming large. Small groups, whose numbers must remain constant—this is achieved through strict sexual taboos and, if necessary, through infanticide—live in scattered pockets throughout vast territories. They are often highly aggressive in nature (the Pygmies are an exception here): for example, the Yanomamö Indians, who bring up their children to be fighters (Chagnon 1966). When the pressures and constraints of the environment deprive human beings of the freedom to develop their personality, they become aggressive, no matter what culture they belong to.

All the great civilizations were the fruit of open landscapes, of the savannas and the great river valleys, not of the forests. What we find in the Earth's forests are relict cultures and displaced peoples who have remained in the primal hunter-gatherer state, with a way of life bound very closely to the environment.

A GIANT ORGANISM IN DECLINE?

The forest in its primeval form has traditionally been regarded by humans as hostile, no matter what part of the world we are talking about. In this Europe was no exception. The rise of its culture, based upon well-ordered agriculture, really only got underway with the clearing of the forests. Although it may be "politically incorrect" to say so, we should not be deceived by the modern city-dwellers' romantic attitudes. There is a strong desire to "go back to nature" and the forest is made the idealized focus of this longing, but the fact remains that the forest, in its natural state, provides no adequate basis for human culture of any scale, and has always been alive with dangers of one kind or another. (This earlier experience is reflected in myths and fairy tales, where the forest is always a hostile place, foreboding and full of demons and wild animals; in the forest you are likely to lose your way.) The clearing of the forest was the legitimate *outward* expression of an *inner* need human beings had to make the landscape their own, to tame the power of wild nature and set their cultural stamp on the land.

This desire lives undiminished to this day in the minds of many members of cultures outside Europe, and is expressed, for instance, in the Neo-European urge to push back the *mato,* as the rainforest is disparagingly known in Brazil (the word means "scrub"; the correct term for forest would be *floresta*). Even scientists who have come to do research on the rainforest have been known to develop such a phobia against it that no power on Earth would induce them to visit it a second time. They have the feeling of being trapped, smothered by its rampant vitality.

That which in former times brought great benefits—the transformation of wild forest into a cultivated landscape—is a form of cultural behavior that has already overstepped the limit beyond which lies the threat of ecological catastrophes. Today the danger is increased by the predominance of additional motivating factors that have already been in play for quite some time. We now have forest destruction pursued solely for profit, which is able to carve its irresponsible path through whole areas where laws and conservation regulations exist only on paper. The legal vacuum concerning the rainforest exerts its fatal pull on plunderers, not only locally, but also worldwide. In the fast-disappearing rainforests of Southeast Asia the destruction is carried out by the big firms that control the tropical hardwood market. They simply flatten the forest, one large area at a time, damaging the host country on two counts—it not only loses the trees, but the profits are creamed off into foreign banks. In Amazonia, where there is little in the way of valuable timber, vast areas of forest are transformed into intensively managed ranches, which bring a handy return to the multinational concerns engaged in this activity. Naturally, the tax incentives for carrying out such "economic development projects" are very generous.

Is the forest inevitably doomed? No, there is hope for it yet, and considerably more than just a glimmer. In order to appreciate the full significance of what hopeful signs there are, however, we first need to look at a further aspect of the rainforest.

It has already been mentioned that in the tropics the herbaceous vegetation that covers the forest floor in temperate latitudes has been lifted up into the forest canopy, and with it numerous animals that "actually" belong on the ground. This tendency for life processes to strain away from the ground, almost to detach themselves from the earth and form their own "realm of light" above ground, is a typical feature of the tropics. It is in no way confined to the rainforest, although here it clearly attains its high point. The opposite picture is presented by the arctic regions, where the vegetation clings to the

Fig. 6. In the tropics the greatest variety of life processes is concentrated in the canopy of the trees, whereas in the moderate zones a corresponding diversity is found immediately above and below the ground.

earth. Both these habits are clearly correlated to the angle of the Sun. By the same token, then, in temperate regions we have "arctic" conditions in winter—the trees become dormant and only the ground vegetation remains green—while in summer we have "tropical" conditions: the herbaceous layer expands upward and the sap rises in the trees.

In the tropics, then, the tree is the dominant form of vegetation. Plant families that are mostly or exclusively herbaceous in temperate climes occur mainly in tree varieties in the tropics (for example, the large pea family, the Leguminosae). And not infrequently those plants that retain their herbaceous habit take the form of lianas, creepers, and epiphytes, using the trees to ensure their "upward mobility."

The purely American Bromeliaceae family (to which pineapple belongs) takes this tendency the furthest: many of its species dispense with roots altogether and instead take up minerals in the form of dust through special organs at the base of the leaves. In this way they are able to live even on telephone wires. Most epiphytes do not go quite so far, however—they live off the humus in the "hanging gardens" that form on the large branches of trees. Many of them are also able to enrich it through special organs on the undersides of their leaves. Many microorganisms that live in the ground in temperate forests are here found between 40 and 60 meters up in this "elevated forest floor." Up there also are the large herbivores—the "grazers" of the treetops, among them one with what approaches a ruminant digestion, the sloth. But in all tropical forests this role is filled chiefly by monkeys: the langurs in South Asia, the colobus species in Africa, the howlers in South America. The forest canopy is also the home of the abundant bird life. The birds share the scene with many species of frogs, which breed their tadpoles in the miniature ponds that form in the rosettes of bromeliads. The pattern is similar for numerous species of ants and termites—they build their nests not on the ground, but up in the very tips of the trees. Some ant species even take the step of carrying certain seeds into their nests, where they germinate, so that before long the ants end up with their own "gardens" (Hölldobler and Wilson 1990).

Whereas in temperate forests, for instance in those of central Europe, the greatest diversity of living processes and of species, especially microorganisms, is found in or just above the ground, in the rainforest this diverse realm is transposed into the canopy. The fact that root systems are sparse and shallow is a telling indication of this. Most roots are above ground, in the form of angled buttresses or stilts, or actually hanging free in the air. The massive runners that coil all over the ground like thick snakes—a typical feature of all the Earth's tropical rainforests—do not, as might be thought, continue underground. They come to an abrupt end just under the surface, branching out into short, pencil-like rootlets, which grow down and branch out in turn into still smaller ones, eventually forming a fine mesh (Fig 7).

Moreover, there is nothing in the ground for the roots to take up; it is barren and contains no minerals of any use to plant life: soluble substances have long since been washed out of the ground by the torrents of rain that lash the forest, day in and day out. This is an astonishing paradox: the Earth's most luxuriant vegetation grows on the most infertile of soils.

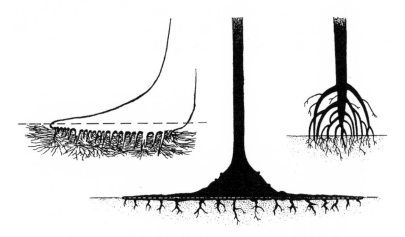

Fig. 7. More on than in the ground, buttress and stilt roots make trees look as though they had been pulled up out of the ground by a giant hand. Left: buttress roots of a kapok tree from which short, cone-shaped extensions protrude into the ground. Growing from these extensions, a dense network of fine roots penetrates the topmost layers of soil. Compare Fig. 4 in chapter 2 above. (From Vareschi 1980.)

Minerals are, of course, present, but they never get into the soil, being totally bound up in the forest's biomass. The nutrient cycles are completely above ground. Leaf litter and deadwood immediately fall prey to fungi, which live in such close symbiotic association with the shallow root systems of the trees that they prevent the nutrient cycles from dipping any lower and feed all decomposition products straight back into the upward nutrient stream of their host plants.

This recycling, however, does not achieve absolute efficiency. Losses through leaching, which over time amount to considerable quantities, are unavoidable. The Amazonian rainforest would have been facing a natural doom if the deficit had not always been made up, and from a very unexpected quarter. The Passat, the wind that blows off the Atlantic from the northeast, brings with it mineral dust from the Sahara (Prospero et al 1981). The Earth's "death pole" is quite literally a key element in keeping the "vegetative pole" alive. This is a truly astonishing example of ecological interdependence. Here two of the Earth's large-scale ecosystems, geographically separate and widely differing in function, are seen to be so attuned to each other that their behavior can only be compared to that of organs within an organism.

This also solves the puzzle of why the trees closest to the sea on the east coast of Brazil are so heavily laden with epiphytes—especially bromeliads, which are rootless and thus totally dependent upon wind-borne minerals. Evidently sea air is not so free of "impurities" as is normally thought. In the case of the southeast coast of Brazil (Fig. 8) it is very likely that the airborne minerals are blown across from the Namib Desert of South-West Africa.

On a number of counts observation leads to the inescapable conclusion that in its present form the unique living system we call the rainforest represents the final state of a long process of development. It cannot have originated on soils poor in nutrients, otherwise it could not have come about at all. This is borne out by the fact that today the forest has almost no chance of regenerating once it has been destroyed. Its ability to fix the mineral nutrients in the living realm above ground can only have developed very gradually as a necessary response to the increasing depletion of the soil. That the rainforest is a very finely

Fig. 8. A tree overgrown with epiphytes on the Brazilian coast between Rio and Santos. The large dark clusters are bromeliads as are also the garlands of Spanish moss *Tillandsia usneoides,* a rootless species. A segmented cactus *Rhipsalis* is seen hanging at left. (From Suchantke 1982.)

diversified ecosystem is seen both in the sheer number of species found there and in the astounding variety of highly complex symbioses between plants and insects (for example, the many plants functionally linked with ants in one way or another, or the highly refined methods of pollination in certain orchids). The number of species is large in the case of trees, but overwhelming in the case of insects. (On the small island of Barro Colorado in the Panama Canal, over 20,000 species of insect were found, about as many as live in the whole of Europe; from ten trees in the rainforest of Borneo 24,000 individuals from 2,800 arthropod species were collected [Reichholf 1990].)

But it is not just that diversification reaches its highest pitch in the rainforest. The opposing tendency toward morphological convergence, that is, the tendency of widely differing organisms to assimilate themselves to a stereotypic pattern, also arrives at a level of perfection not found anywhere else on Earth.

Development and maturity, however, could at a certain level also mean overmaturity and aging. There is indeed some evidence for this. A comparison with a completely different ecosystem might help to make this clear. If the climax of plant life on land is found in the rainforests of the equatorial belt, the most productive regions in the oceans lie close to the poles. These "algal meadows" have only a fifth of the amount of chlorophyll found in a comparable area of rainforest, but in spite of this their assimilation rate—the amount of oxygen given off into the atmosphere per unit of chlorophyll—is five times that of the rainforest (making the algal flora of the oceans the most important, almost the sole source of atmospheric oxygen) (Reichholf 1990).

Here lies one of the keys to understanding the rainforest. While the tiny, short-lived algae never build up any kind of appreciable mass and live in a constant dynamic *flux* of growth and decay, the life processes of the rainforest are more than correspondingly slow and have a tendency to bring the substances they take up to a *standstill*, to fix and store them. The result is the great accumulation of biomass found on land. Let us remind ourselves of the figures quoted earlier: although the rainforest covers only 3% of the Earth's land surface, it accounts for almost a third of the total biomass—just as much as that in all the oceans, which together take up 65% of the Earth's surface!

The crucial point here is that only a small proportion of the rainforest's biomass is truly living substance. Of the 1,000 tons per hectare, only 20 to 30 tons are living leaf-matter. By far the most part is dead wood, in other words, carbon in fixed form—and what a form! If we consider the rainforest from the point of view of the tremendous density and ironlike hardness of its timber—timber that sinks in water like a heavy stone—it appears in a rather strange light: it displays an extremely strong tendency toward mineralization, stronger at any rate than in any other living system, and on a scale that is outstripped only by coal formation. This affinity with coal formation appears nowhere so strikingly as in ebony, the black color being a clear indication of its high carbon content.

That this mineralization process stops short of coal formation is due entirely to the combined efforts of the teeming hosts of decomposers and compost-making organisms in the forest: the fungi, but also the ants (the actual lords of the forest, occurring in truly vast numbers and in a very wide range of species) and termites. These efforts include cooperation between the two groups of organisms, for example in the underground fungal gardens of the million-strong colonies of leaf-cutter ants.

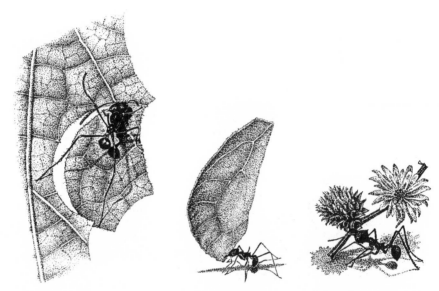

Fig. 9. Composters of the rainforest: Leaf-cutting ants harvesting parts of plants; these are stored in subterranean greenhouses where they are inoculated with fungi whose spore cases serve as food for the ants. The rotting remains enrich the soil with humus. (From Suchantke 1982.)

Viewed in the context of marine vegetation, the earliest, one might say "youthful," phase of the plant world, the rainforest appears as the most highly differentiated and most aged stage in the development of the Earth's vegetation. This impression of old age is strengthened when the transferal of the forest's living processes from the ground into the canopy is added to the picture. Something that originally was closely bound up with the earth and permeated its upper layers has successively withdrawn. In the large-scale organism of the rainforest, something is occurring that is the same as the process all organisms go through when they die: the life processes withdraw, they "excarnate."

It is no wonder, then, that all it takes is a little nudge, and the rainforest is destroyed forever—a state of affairs quite unlike that in temperate latitudes, where living processes are (still!) so connected to the earth that if the landscape were left to itself the forest would immediately return.

IS REJUVENATION POSSIBLE?

What has been said up to now should in no way be misunderstood as a plea for a *laissez faire* attitude to the continuing destruction of the rainforest, based on the motto, What does it matter, we're only following nature by helping a dying ecosystem on its inevitable path to oblivion! Nothing could be further from the truth. The rainforest is not about to die tomorrow—we can expect it to last for many millennia yet.

The point is, rather, that the human factor in relation to the rainforest need not be purely destructive. The conventional attitude sees conservation as the only alternative to destruction, but what is required here is something more radical and far-reaching. Between the two extremes of wanton destruction and absolute protection there is another possibility: *rejuvenation.* This would involve intervening in the senescent stasis of the rainforest ecosystem in a way that would return it to a more "youthful" stage. This could mean introducing something new into the system or releasing the forest's own powers of rejuvenation, but either way it gives humans a creative role, which demands a great deal of effort and commitment. Utopian as this may sound, it is entirely practicable; and it is extremely urgent, for we are in a race against time, against the ever-encroaching destruction that is proceeding by giant steps and will soon attain a momentum that will prove unstoppable.

History provides us with ample evidence that humankind is capable of bringing about the rejuvenation of mature ecosystems that have reached a certain end-state (a state in which no further change will take place, assuming the climate stays constant). In a wide variety of locations over the whole Earth the original natural landscape has been transformed into one shaped by agriculture and the patterns of human settlement. In Europe, before humans began to exert this cultural influence, a deciduous (mostly beech) forest, exceedingly poor in plant and animal species, had spread over large areas. Clearing it, of course, meant destruction—but that was by no means the end of the story. Human culture eventually replaced the forest with a newly created and richly structured landscape of meadows, fields, orchards, hedges, copses, and woods, which provided habitats for a wide variety of wild plants and animals that had not been present before. To these were added the many new breeds produced by animal husbandry. Today only remnants of this unique synthesis of nature and culture exist, but new ecological criteria have taught us to appreciate it anew, and there are definite signs that it is on a comeback. Such a reshaping of the landscape certainly involves radical intervention in nature—but in this nature is not diminished. Quite the reverse, in fact—nature undergoes further development. Left to itself, nature would have maintained its achieved climax state indefinitely, with at most tiny variations from time to time (Suchantke 1990).

If this sounds too optimistic or too much like wishful thinking, we should bear in mind that nature itself does not behave any differently. Two opposing tendencies are constantly at work within it. One moves toward invariance, stability, and balance, ensuring that organisms are perfectly adapted to their environment—in short, toward a steady state. The other moves toward drastic changes, usually occurring in the form of *external* attacks—climatic changes, floods, continental drift, mountain formation, etc.—which mean catastrophe for the steady state and lead to mass extinctions. At the same time, however, the changes provoke new developments—new landscapes with new ecosystems emerge, new animal and plant species spread over the land.

It must be admitted, of course, that the conditions that render European beech forests susceptible to transformation into fertile agricultural land are not those that obtain in the rainforest. This means that methods appropriate to the one region cannot be applied to the other. To do so would simply be propagating ecological imperialism (or colonialism) in a modern guise. Each specific situation demands specific methods, developed locally to establish the conditions that in more temperate climes such as Europe are given from the outset—in other words, to *create a fertile soil.* The cycles of life, which hold the key to fertility and which have been so rigorously withdrawn from the earth in the rainforest, must be anchored in it anew, must be, as it were, "reincarnated."

This almost sounds as if the methods have still to be developed. In reality they have existed for a long time, but they have been ignored. For why should anyone have paid any attention to the methods of "wild," "primitive" peoples—especially when exploitation, and not care or cultivation, was the order of the day? Only very recently, since a change in attitude began to make itself felt, have the astonishingly refined methods of agroforestry practiced by numerous peoples in all tropical countries come to light—and that not a moment too soon, for many of these peoples have themselves fallen victim to the same forces that threaten the forests (Gómez-Pompa and Kaus 1990).

The Kayapó Indians of the Amazonian rainforest are one of the most impressive examples (*WWF News* 1987). In contrast to the other tribes ranging through the region, most of whom are hunter-gatherers, they pursue a highly developed form of forest cultivation, which goes hand in hand with careful conservation. For instance, the Kayapó, who were recently studied by a team of researchers, plant tree seedlings all year round, especially Brazil-nut seedlings. It had formerly been assumed that Brazil-nut trees were all wild and not able to be culti-vated (they grow extremely slowly into rainforest giants, and all the Brazil nuts we buy over the counter have been collected in the forest). Another of the Kayapó's practices is the occasional felling of single trees, which creates enclosed sunlit clearings where hundreds of med-icinal herbs either are planted or seed themselves. In addition a luxu-riant ground cover emerges, which is absent elsewhere and attracts game. In island-like clearings, the Kayapó manage, through highly refined composting techniques and the selective addition of certain plant ashes (they can distinguish dozens of different sorts), to create fertile loam—an astonishing achievement considering that the untreated soil is as good as sterile. Crop plants are sown in mixed cultures made up of a large variety of selected species. The Kayapó call these *ombiqua o-toro,* which roughly means "friends in growth"—an accurate description, for the plants do grow better in these combinations than they would alone. "This could revolutionize the methods by which agronomists construct and plant fields, and take the place of unstable monocultures," said the leader of the research group, Darrel Posey of the Museum Paraense Emilio Goeldi in Belém.

Investigations in Mexico found it highly probable that the extensive lowland forests (for example, on the Yucatan Peninsula), if not entirely anthropogenic, owed their form and species composition not merely to indirect human influence, but to direct human action (Gómez-Pompa and Kaus 1990). During the period of the Olmecan and Mayan cultures, long before Spanish colonization, these regions had a much higher population density than today. Accounts from the conquistadors bear witness that in their time the forests were under intensive use and for millennia had probably played a vital role in providing the people with food and resources. Even today the time-honored techniques are still in use wherever the people have not become estranged from their cultural roots. Certain species of trees are protected and nurtured; others eliminated and replaced by desirable new ones. All this is done in a way that preserves the natural structure of the forest, and the "artificial forest" can be distinguished from the original one only by the species composition: the tree species used most by humans dominate the former. Gómez-Pompa and Kaus (1990) list 23 species with edible fruits, 36 that have medicinal uses, and 37 that provide wood and other craft materials. (Just how valuable the products of these trees are is demonstrated by one species, whose fruit, the avocado [*ahuakatl* in Aztec], is now widely popular and is one of the most nutritious fruits, rich in protein, unsaturated fats, minerals, and vitamins [Brücher 1977]).

There are, of course, agroforestry approaches of modern origin. They are all unconventional, pioneering enterprises, but they are standing the test of time. Some modern agroforesters follow a method similar to that of the Mayan farmers, in that they only thin out the forest a little and sow their crop plants in a way that preserves its natural layered structure. Others—and their achievements cannot be praised too highly—have succeeded in restoring fertility to leached-out, sterile soils. Foremost among these are the Estancia Demetria at Botucatu in the state of Sao Paulo (Suchantke 1987) and Granja Imperial in the Amazon Delta region (Sioli 1973), but other enterprises in the Dominican Republic (Samwald and Ditfurth 1989) and Mexico (Neumann 1979) are also worthy of mention. As varied as their approaches are, they are united in being unconventional—either organic or biodynamic. One of the keys to their success—to name just one example from the rich diversity of methods they use—is treating the soil with the right kind of manure. Organic compost, when used correctly, is resistant to leaching, unlike chemical fertilizer.

Organic agriculture is proving its worth to a more striking degree in the tropics than in temperate regions. With its ability to come to terms with the fragile ecosystems and worn-out soils found there, it may indeed be the only way of giving agriculture a viable future in the tropics. Nature also stands to benefit from it. Through its rejection of monoculture and its creation of a diverse pattern of small cultivated plots of different kinds—fields, meadows, groves—it creates a rich mosaic of biotopes and habitats in which numerous wild plants and animals can find a home.

If such ecologically based agriculture were to spread, it would bring with it other advantages. In contrast to the giant monocultures of the plantations, which are tended by machines or migrant workers and have caused the depopulation of large areas in the South American interior, organic agriculture is very labor intensive. Many people forced to join the never-ending stream of migrants to the slums of the big cities could be resettled and could make their living on organic farms. Such resettlement will in any case have to take place sooner or later if a country like Brazil is not to collapse in hunger and social catastrophe.

This prospect, however, is fraught with difficulty. Apart from a few perceptive individuals with a feeling of responsibility, there is no one who can even see the problems, let alone what needs to be done; and many who could see them avert their eyes out of indifference or egotism, or both. The perceptive few live in the cities, an ineffectual minority of intellectuals and scientists, while those involved in the front line are either blind to the ecological consequences of their actions or lacking in good will. Lack of good will, however, is by no means the whole story. The majority of the population—half-castes and Neo-European immigrants who arrived generations ago—lead a lonely existence in the wilds in a state of sometimes extreme cultural impoverishment. They are largely illiterate, and their worldview is a mixture of elements from Christianity and animism (Gerbert 1970; Suchantke 1987). Everything that happens is explained as the actions of either spirits or demons. Nothing in their social background provides them with the qualities of independence of mind or the ability to act on their own initiative. They stand, as it were, in a cultural no-man's-land, cut off from the strong oral culture of the past and having at best a tenuous relation to the literate culture that replaced it. Thus any kind of new beginning here will depend upon educating a population for whom ecological thinking and careful treatment of nature are still unknown territory. This educational task is as important as the already pressing one of agriculture, if not more important.

For the Western world there are, of course, lessons to be learned here. We could begin, for instance, by correcting the false assumption that preservation and cultivation of the rainforest are totally incompatible. Unfortunately the media, as well as practically all the books on the rainforest currently flooding the market, continue to promote this assumption. It is clearly apparent that some of these books have been written by biologists who would dearly love to exclude all humans from the rainforest (apart from a few highly qualified specialists) in order, at all costs, to avoid any disturbance of its biodiversity. They would like to turn the rainforest into a museum. When set against harsh reality, this extreme, purist version of conservation is soon seen for the impracticable illusion it is. It holds no promise for the future and is only concerned with preserving the legacy of the past. Moreover, its determination to keep people out—not only their destructive tendencies but also their creative potential—makes it antisocial and undemocratic.

The rainforest is very popular in the Western media. People react to its plight with a mixture of shock, nostalgia, and resignation: it is tacitly assumed that all attempts to save it are bound to come to nothing. Promising new approaches and time-honored traditional methods, which leave no doubt that its salvation is possible, remain unknown, hidden away in scientific journals. They are simply not spectacular enough, and cannot compete with the horror reports at grabbing headlines.

Nevertheless, the author of these lines has experienced time and again when speaking about these things before audiences of students that as soon as the subject of the Kayapó or the agroforestry methods of the Mexican Indians or the remarkable achievements of Estancia Demetria is broached, there is a spontaneous surge of enthusiasm, with not a few of the young audience expressing the wish to be there doing their bit. If these things could become general knowledge, international pressure upon the countries concerned would no longer need to consist merely of measures to force them to halt the destruction, but could be increasingly aimed at encouraging existing and genuinely workable alternatives. That would be developmental aid worthy of the name. And it would have no difficulty being accepted by the recipient populations. For one thing, it would involve no threat of exclusion from protected territories, but would instead offer the possibility of a better standard of living and the promise of a dignified future. For another thing, such alternatives would not simply be yet another form of colonialism, albeit in a new "ecological" guise, but would be based on the adoption and further development of local methods. A. B. Anderson has edited a book (1990) in which he has collected and described a range of alternatives to rainforest destruction. His words at the beginning of that book may serve as a fitting summary to conclude this chapter: "Given the complex situation described it would appear that the complete destruction of the Amazon rainforest is only a matter of time. Yet this text is based on the firm conviction that the current scenario *is* reversible, and that the viable alternatives to deforestation—both within Amazonia and elsewhere in the moist tropics— *do* exist."

Humankind and Nature in Different Cultures and Continents

THERE IS NOWADAYS a widespread conviction that human beings have always played the part of destroyers in their dealings with nature. They could not have acted otherwise, it is thought, since humans are inherently destructive, and if we want evidence we need only look at history. Such convictions are heard most readily from the lips of Western technocrats and are applied as a matter of course in pronouncements about other cultures and cultural epochs. The fact that at certain times and places nature has been treated intuitively in a more caring and more ecologically sound way is simply ignored. Instead these technocrats tend to castigate the achievements of other cultures as backward or, at best, "prescientific." In this they are guilty of chauvinism, even cultural nihilism, and although they may take a "politically correct" line on the deleterious effects of economic imperialism on native cultures they nevertheless remain blind to the fact that their own arrogant attitude is simply another form of it. Moreover, by eliminating all alternatives, this blanket interpretation of the human-nature relationship becomes self-justifying and self-perpetuating, styling itself as the only way to "get a grip" on the situation, the sole alternative being resignation.

Something approaching the true complexity of the relationships involved in culture can be revealed by comparing two widely differing regions. Here our gaze will be turned, on the one hand, on a region of Southern Asia where cultural practices inherited from the ancient past are still very much alive and have a determining influence; here relationships that are in no way hostile to nature prevail, and nature is experienced not as an "environment" lying outside the realm of normal cultural life, but rather as a sphere in which culture is enclosed, so that separation of the one from the other is unknown. In contrast to this we will look at a region of the Earth that was never molded by any higher culture before representatives of modern materialist civilization invaded it; it thus had nothing in the way of traditional, autochthonous cultural elements to set against the imported technologies and the attitudes they embodied. The region referred to is South America or, more exactly, the extra-Andean, non-Incan part of this continent. Let us consider it first.

SOUTH AMERICA: An Impoverished Land of Homeless People

Although they have been there for centuries, this continent's colonizers and immigrants have not yet fully made it their home. They have succeeded neither in creating a culture with its own distinct identity (though varied and significant beginnings have been made), nor in developing a viable relationship to the natural environment of the region. The caring and ecologically appropriate methods of cultivation and economic management that might be associated with such a relationship are similarly lacking. No matter what the immigrants themselves might think, they cannot be regarded as bringers of culture. On the contrary, they are creating a state of affairs in which the elimination of the remnants of their own imported European culture is progressing so rapidly that the calls for renewal are getting louder every day.[1]

---

1. Symptomatic both of the decay of the colonial culture and the search for novelty are such phenomena as the increasing rejection of Catholicism and the embrace of imported and syncretically altered African forms of religion, and of magic, spiritism, and Protestant revivalist sects, as well as the fact that there are adherents of folk religions spread widely through all social classes (Gerbert 1970).

This may well have to do with the fact that the very earliest settlers had already gone a long way toward breaking the ties of nature in their countries of origin. At the beginning of the modern age, the focal point of cultural development shifted from the country to the town. With that, human existence was no longer embedded in the necessities of natural processes as it had been not only for farmers and peasants but also for the agriculturally based monasteries of the Middle Ages. Trade, science, and mechanical inventions gave a new edge to the ethos of the city that cut through the bounds of nature, favoring the unfolding of the autonomous personality—emancipated, free, but also rootless and homeless. This is the birth of the discoverers and colonizers. No new place can provide a home for them, for they take the spiritual guidelines of their existence, if at all, from their own inner life and not from interaction with the natural landscape of their homeland. Hence the striking lack of fit between South American cities and their natural surroundings. The landscapes in which they have been so incongruously set are either still in the their original wild state or else already destroyed, but scarcely ever are they permeated with human culture.

Of course this sense of being at odds with the local landscape is just as evident in Europe. It is simply masked there by the relics of nature-related culture, which persist in many places and further disguise the fact that the making of the landscape is a cultural achievement belonging indisputably to the past. Significantly enough, there is one group among the new citizens of South America to whom these observations do not apply to the same extent: the (forced) immigrants from Black Africa. It seems that out of their basic religious reverence for nature they found it much easier to make a home of the familiar world of the tropics.

It is not just nature, however, that suffers a lack of appreciation; the general attitude to the indigenous peoples is no better. Indeed the idea that they are something scarcely human has persisted right up to the present day. Until recently they have been regarded as subhuman savages. In all southern continents—Australia and Tasmania, Africa, South America—the same picture is found. Initially the people of these indigenous cultures, who in some respects are still at a more youthful, childlike stage in their development, many of them hunter-gatherers still to take the step toward becoming transformers of nature, invested very high hopes in the newly arrived Europeans. Laurens van der Post (1955) describes this feeling, as it appeared among the African peoples, in the following terms:

> I think the initial readiness of the black man to serve the white man was perhaps because, unconsciously, he had long waited for someone like the white man to come and bring him something which only the white man could provide. When he did come, it was as if it was in answer to some dream far back in the African mind, and in response to some deep submerged hope that Africa had of the future. The white man's coming seemed to imply the fulfillment of a promise which had been made to the African far back by life in its first beginnings.

Similar expectations were to be found across the Atlantic—among the Central American cultures there were even actual prophecies that seemed in the eyes of the Indians to have been fulfilled with the arrival of the white man. And in South America, in the inner reaches of Brazil, time and again throughout the centuries migratory waves of messianic enthusiasm have set whole tribes upon pilgrimage to the coast, seeking redemption. A pessimistic mythology foreseeing the imminent end of the world drove them to seek the "land without evil" lying, so they believed, in the east, from where the Europeans came (Gerbert 1970).

But the meeting, the exchange, the restorative enhancement, did not take place. The white man remained apart, so that what ultimately lay behind the great urge toward discovery and colonization was denied expression. The whole enterprise can be seen solely in terms of lust for power and material possession—but that was only its outward side. Nevertheless it succeeded very well in masking what was going on below the surface. To quote Laurens van der Post again, a passage relating to Africa, but nonetheless of general validity:

> The more I learn of so-called primitive man, the higher is my regard for him, and the more I become aware just how much we could learn

from him and what profound insights he could impart to us. I believe
he is just as necessary for us as we are for him. I regard us as two
mutually enhancing halves, marked out by life to form a whole. The
longer I contemplate the ever-darkening stage of contemporary
events, the more I realise how vital our need of each other is.

Described here is what should have been the first steps toward a
culture encompassing the whole of humankind; one that would lead
neither to one culture suppressing and eliminating all the others, nor
to a general leveling out with concomitant loss of individual identity,
but rather to a worldwide diversity. Within this each culture could
make its particular contribution and have its due place, and together
they would not merely complement each other, but further development, mutual inspiration, and continual learning would occur.

So far there are only a few modest attempts at this in evidence.
The black African has been bitterly disappointed and left in the lurch
by the white man. It may be, however, that events now taking shape
in certain parts of Africa are a sign that the black man has not given
up and out of his own insight will force the unwilling whites toward
brotherhood. The first beginnings of such an awakening are already
detectable.

The case of the Indians is different. They recoil from contact with
the modern immigrant civilization, one touch of which is enough to
threaten their entire existence. They have no place in it, and if they
are paid any heed at all it is mostly as an obstacle in the way of the
relentless expropriation of land. The Indians do not, or, rather, should
not exist! This attitude is exemplified by the fact that the Indian population of Peru—the representatives of a high culture stretching back
thousands of years—were granted full citizens' rights only in the
1920s, and that largely only on paper. Further evidence—perhaps not
so weighty, but nonetheless symptomatic—is provided by the basic
local geography textbook used in Brazilian schools, *Tipos e Aspectos do
Brazil*. Here the children are introduced to the diversity of the country's landscapes, their fauna and flora and the people who live in them.
They are told about the fishermen of the coasts, about the *Caboclos*

Amazonicos, the smallholders colonizing the rainforest, about life in the arid Sertao region in the northeast, the tea-pickers and gauchos of the south; even the *Favelas* of the big cities are mentioned. Of the Indians there is no hint, in either word or image; they are simply not part of the official picture. Between 1957 and 1975 two hundred and thirty of the existing tribes were lost (Goodland and Irwin 1975). How many have disappeared since then?

The white South Americans' attitude to nature is just the same. They perceive the wild environment around them as only a chaotic and hostile world, which must be resisted, driven back, destroyed. In Brazil the rainforest is usually referred to as *mato,* which means something like "bush" or "scrub" (the proper word for it would be *floresta*). One richly forested Brazilian state even carries the name *Mato Grosso.*

Fig. 1. The burning forests of Brazil.

To burn, to destroy this useless *mato* is regarded well-nigh as a duty. Any bits of forest still left beside major highways, for instance, are at the mercy of roadside picnickers, who are very likely to set it on fire— just for fun. On weekends this is a common sight in Brazil.

Nature, for her part, and this is the other side of the story, does almost nothing to accommodate humans. The rainforest, the climax and most resplendent manifestation of plant life on Earth, is rooted for the most part in completely infertile, agriculturally worthless soils. This is an apparent paradox: surely where plant life grows in such profuse abundance the soil should be exceptionally fertile. The explanation is that the tropical flora, having extracted, over eons, the life-essential minerals from the formerly fertile soil, now retains them in cycles which no longer include the soil, for no sooner do they reach the ground than they are taken up by surface-creeping roots.[2]

The soil conditions of the *Cerrado,* a type of vegetation for which the terms *bush* and *scrub* are much more appropriate than for the rainforest, are even more extreme. It is found on the endless plains of the interior and east of Brazil, sometimes in the form of thickly tangled scrub (*Cerradão*), sometimes as open grassland with single small trees (*Campo cerrado*). The landscape looks exactly the same for hundreds of miles. It is gray or yellowish-brown, bleak. The flora gives the impression of being stunted and impoverished. The twisted shapes of the small trees with their tough bark and thick branches—thick by virtue of their sparseness—look as if they are having difficulty raising themselves above the ground. This impression is thoroughly accurate, for many of the species found here also occur in the forest regions, where they manage to grow to a considerable height. Here, however, they remain dwarfish and deformed. The tops of palm trees sit on the ground, seemingly without trunks, but they do have trunks—underground. Also many other plant species that remain small above ground

---

2.   In the essay "What Do Rainforest Have to Do with Us?" in this volume this phenomenon is dealt with in more detail. A still more detailed account is to be found in the chapters *"Die Amazonas-Niederung als harmonischer Organismus"* and *"Die brennenden Wilder Brasiliens"* in the author's book *Der Kontinent der Kolibris. Landschaften und Lebensformen in den Tropen Sudamerikas,* 1982.

will be found to have thick lignified, tumor-like roots with offshoots of disproportionate size—the so-called xylopodia, strange structures serving no storage function. All these morphological peculiarities are clear evidence of senescence, or ageing, a process associated with life in ancient, largely demineralized soils, as is the case here. In coping with these conditions the flora is aided by an excess of growth-inhibiting aluminium.

In a bush and savanna landscape such as this it is also hardly surprising that there is a great scarcity of animal life—that is, apart from birds. Of these there is no shortage: hummingbirds, toucans, and, at least at one time, macaws with their gorgeous plumage. But with mammals it is a different story. In landscapes in Africa that are outwardly very similar to this one, herds of grazing animals and groups of large predators dominate the scene to such an extent that the air is charged with animate life and the observer is gripped, overcome by its compelling power, but in the open plains of South America one's attention will be caught, at most, by the leaping of little pampas deer or duiker-sized brocket deer, or by foxes or small cats.

Humans have as yet had little idea of what to do with these vast expanses of impoverished nature, which are as little suited to agriculture as the rainforests. Nevertheless in the course of their cultural expansion humans have, albeit involuntarily, enlarged such areas. This does not mean, however, that the original Cerrado has been extended. Rather it has led to the formation of a Cerrado-like secondary landscape, which differs from the original by being even more barren. In many places in the east of Brazil, in areas near the sea and in river basins where the tropical forest used to stand, the soils have been reduced by rapacious agricultural practices and above all by coffee plantations to such a poor state that only a barren plain is left, a landscape of stunted bushes and termite mounds where the grass is too tough for cattle to eat.

The one thing there is really no deficiency of in South America is depressing evidence of the skewed relationship, even the absence of relationship, between humans and nature.

SRI LANKA: Original Oneness of Culture and Nature

The images encountered in Southern Asia are quite different, at any rate in those regions where Buddhist culture still thrives, as, for instance, in Sri Lanka. They are certainly worlds away from what we experience in South America.

Our first impression, and one that proves ever more justified, is that there is no separation, certainly not a trace of hostility, between culture and nature. Where the one stops and the other begins is impossible to determine. Characteristic of the Sri Lankan landscape are the many pools and lakes covered with floating tapestries of water lilies, out of which tender pink lotus blossoms rise on tall stems—and every one of these pools is artificial. They were made to serve as reservoirs for the paddy fields, and their origin goes back partly to pre-Christian times.

For animals they are a true paradise. A two-meter-long Indian monitor lizard suns itself on the bank, looking like some prehistoric dragon. At our approach it shows no sign of quitting its post, being well used to the presence of humans, who value it as a killer of snakes. Another dragon, easily twice as long, dozes across the way on a mudbank, its mouth wide open: a mugger crocodile, easily mistaken for one of its African cousins.

Where the water lilies form a floating meadow the inhabitants are much more graceful. Black and white pheasant-tailed jacanas, slender and elegant, with lightly curved tail-feathers, pick their way carefully over the leaves on toes specially lengthened for the purpose. An iridescent purple gallinule—a hen-sized relation of the coot, but with a red rather than a white frontal shield—has a lotus bud clamped between its coral-pink toes and pecks it away bit by bit. Small olive-brown pond herons stand around motionless, looking like wooden posts jutting out of the water—at least until they soar off, revealing their snow-white wings. Kingfishers flash like iridescent arrows over the water, and beside us a male weaverbird proclaims with loud sparrow-like chirps the completion of his elaborate retort-shaped nest, which he hopes will find approval with the best of the neighboring females.

Fig. 2. The delicate grace of the "small white Water-princess" is somewhat diminished by her very large feet. But this makes it possible for her to spread her weight widely on the wavering blue carpet of the lotus meadows.

Some children come tripping along and try to sell us bunches of water hyacinths. Astonished, we ask why we should buy them when they are growing on the bank beside our feet. The little girls laugh— and make us a present of them.

It is oppressively hot and close, and we are nearly perishing with thirst, so we turn to one of the little farmsteads that lie hidden beneath the fronds of the coconut and date palms behind the breakwater. We ask if we can buy a few coconuts, and the obliging man shows us how you can top the green fruits in such a way that a little hole appears, through which the clear cool liquid inside can be drunk with the aid of a straw—without one if you are very thirsty. The whole family, meanwhile, stands around us. One little daughter has on her arm a tame young mongoose, which at once begins climbing all over her. He is both a playful pet and a useful weapon against snakes; the perilous cobra is fond of human settlements for the abundance of hiding places and rats they provide. And the lady of the house shows us with evident pride a tame hind that is also a member of the family.

There is a harmonious relationship between human beings and the creatures of nature that is taken for granted—a sense of kinship that proceeds from humans and is answered by nature with confidence and trust. When, for instance, you try to make a picnic of your midday break in a forest clearing, you very quickly have uninvited guests— palm squirrels, house crows, bonnet macaques—who expect a share and make sure they get it. The atmosphere is one of peaceful tolerance and loving regard for life; it is the spirit of Buddhism, become such a basic habit of life that it permeates not only the human sphere, but also the natural. This brings a heartening realization: nature answers like for like—as she is approached, so she responds. She is manifestly receptive to the moral quality of the manner in which she is treated. It really works! Human beings, so this experience teaches, are not irrevocably destined to be destroyers of nature. It is possible for them to act differently, as long as their actions are based on a sound attitude, be it Buddhist or some other.

The highest expression of this harmonious fusion of culture and nature is found embodied here in a landscape feature whose elements have been combined to form something unique, to which the Occident has nothing remotely comparable: the shrine of the Buddha at Gal Vihara near Polonnaruwa. In gentle meadowlands set with trees and shrubs a group of rocks have formed a natural amphitheater. It gives a feeling of spaciousness and is in no way closed off from its surroundings. Out of one of its sheer walls a group of colossal sculptures have been cut. Among the Buddha figures one stands out through the power of its expression and the extraordinary dynamism of its form: Ananda, the Buddha's favorite pupil. The perfection of this sculpture is due to its being totally in harmony with itself. The poised, suspended vigor of its animated surface expresses inner activity—movement within stillness—which is intensified in the play of the facial expression. The gaze is directed inward in a state of active rest.

Fig. 3. Statue of Ananda, the favored disciple of Buddha, nearly 7 meters high, carved out of a cliff in the twelfth century A.D., Gal Vihara near Polonnaruwa.

A young couple approaches the shrine; the girl is swathed in a gorgeous sari and holds a little bowl of strongly scented frangipani blossoms, which she lays before the statue of Ananda. For a long time the two kneel deep in prayer before the image. As silently as we can we withdraw, so as not to disturb them. But our consideration turns out to have been unnecessary, for the pair are impervious to any disturbance, even when a troop of boisterous children suddenly appear and deposit their offerings of blossoms rather more gaily than devoutly.

And all the while the air, bathed in bright sunlight, vibrates with birdsong, the measures of the sweetest of all singers: shamas and magpie robins. A flock of emerald-green, pink-headed parakeets screeches past. Large elegant long-limbed langurs, the monkeys sacred to the God Hanuman, their black faces framed by hoods of white, sit indolently among the bright foliage. And in a pool a little distance away the white lotus blossoms display their dazzling radiance.

The crowning glories of creation, nature and the human spirit, here join in celebratory union. The marriage of the two engenders perfection, for the one provides what the other lacks: nature replete with radiant light and melodious sound, enchanting the senses and beckoning toward the realm of dreams; her polar complement the wakeful soul, turned inward in deep contemplation of its own inner worlds.

The sculptures of Polonnaruwa stem from the twelfth century, but the Buddhist culture of Sri Lanka is much older. Reaching back to the third century B.C., it was brought to the island by Mahinda, the son of the great Indian emperor Ashoka. There is a legend about this event that contains, like a quintessence, all that still lives today in Sri Lanka. Mahinda, after his arrival, wandered over the island in the saffron robe of a begging monk. One day, in a forest clearing, he happened to meet the ruler of the realm who, with a great train of followers, was out hunting. The king had just pulled back on his bow and aimed at a great stag when Mahinda stepped between him and his quarry and bade him stay his hand. Angered at the brazenness of this stranger, the king demanded that Mahinda explain himself, whereupon he answered by expounding the teachings of the Buddha. The sovereign, deeply moved, took Mahinda into his court and soon after was

converted and introduced the teachings of the Buddha among his people. In Mihantale near the royal city of Anuradhapura, to which
Mahinda withdrew in his old age, stand to this day, beside many other
treasures, two ancient weathered stone tablets, on which are inscribed
the world's oldest nature-conservation laws.

Nature—this much we learn—responds to humankind in the same
measure as she is approached. It is plain that she reacts especially sen-
sitively to the moral-ethical quality of humanity's treatment of her, perhaps because she herself has no knowledge of such categories and is
therefore predisposed to accept those provided for her, as if she had
always expected them. It is therefore no coincidence that in Buddhist
culture there are two domestic animals whose like is not to be found
anywhere else. It is not quite accurate to designate them as domestic
animals, for these creatures are both wild and domestic, without being
entirely one or the other. One of them, the Asian elephant, is a wild
animal perpetually capable of becoming domestic, while for the other,
the buffalo, the boundaries are completely fluid and both states easily
alternate. It is not known whether the populations of wild buffalo, the
arna, are actually purebloods or have been mixed with domestic buffalo from time immemorial. Domestic buffalo are forever running away
and mixing with the wild herds, and having their cows served by wild
bulls is a widespread practice among farmers. According to biologist
Fred Kurt (1986), in southern Sri Lanka,

> the buffalo breeders only keep the animals temporarily in domestic-
> ity for the purposes of milk production. To this end they are placed
> under an irresistable compulsion. This is how it is done: a calf
> having been born in the wild, a group of daring youths will waylay
> it in its hide-out in the undergrowth and carry it off to the village.
> The mother will try to drive the kidnappers away, but they are well
> skilled in heading her off. She grows tired and finally follows her tor-
> mentors for good or ill. She is then shackled and put in a shed. Here
> she has to stay, giving milk both to her human companions and her
> calf. Once it is weaned she will be set free. Mostly the calf rejoins the
> herd as well.

Elephants too keep being transferred one at a time from the wild to the domestic state. Their evident ability to change, as it were, in one step from the natural to the cultural domain seems to point to some essential kinship they have with humans, a point we will come back to later. To breed them in captivity, what with their almost two-year gestation period and still longer immature phase, during which they are unfit for working, would be senseless and uneconomical (although some mountain tribes in Southeast Asia are known to have bred elephants and used the young ones, with their speed and agility, for riding) (Kurt 1986, McKay 1973).

Newly captured elephants are tamed initially by extremely brutal treatment, using force to break their wildness and shyness. But the animal is then appeased by its new master, the mahout, who soothes it with endearments and tender cajoling. From the very beginning its already tamed comrades assist in the process of habituation. Without their help it would be impossible. Certainly the highly developed and in many respects unique social life of these animals paves the way for their "acculturation." Indeed, their social life shows inchoate signs of

Fig. 4. Wild water buffalo and cattle egret, Yala National Park, southern Sri Lanka.

Fig. 5. One of the few places outside zoological gardens where captive elephants have been bred is at Chitwan National Park in Nepal.

true culture. It is characterized by ingrained patterns of mutual aid and protection, as shown by the females acting as midwives and wet nurses for each other, as well as by the lifelong bond between female offspring and their mothers (the males keep more apart the older they get, eventually only putting in brief appearances among the herds). As in all animal communities there is a dominance hierarchy, but it is not fixed and determined by bodily size, as is usual; rather the mothers with the newborn and youngest calves are automatically placed in the highest position, "held in the highest esteem." Just how far the mutual interaction within the communal life of these animals reaches is illustrated by the astonishing fact that the duration of pregnancy is flexible enough to be varied between seventeen and twenty-four months, and can be regulated so that all the births in the herd occur almost simultaneously.

It is no coincidence that it is the Asian and not the African elephant that shows this closeness to humans. In Africa the elephant has never been domesticated in spite of the long pastoral tradition there. The reason for this may lie with the people themselves, for, although there is a vast abundance of animals, domestication of native species has never succeeded. But not only that: Experiments done by Belgian investigators in their former colony of the Congo with the aid of experienced Indian mahouts showed that mature African elephants are not susceptible to taming. The researchers therefore tried catching young ones, but could not get anywhere near them except by shooting their mothers (Hediger 1950).

If the two elephant species—which are not very closely related—are compared, it is readily apparent that the African one is morphologically more perfect and therefore "more fully an animal" than the other. The Asian one, by contrast, appears more juvenile in its proportions. The skull with its high "baby forehead" is rounder, more domelike, and the head is larger and more clearly set apart from the rump than that of its African cousin. The latter, with its receding forehead, whose contours flow without transition into those of the back,

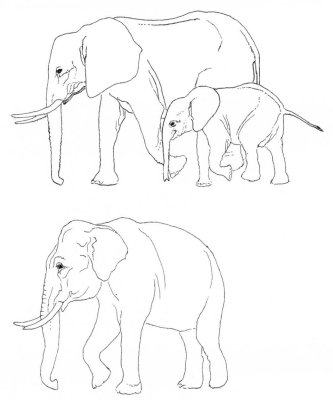

Fig. 6. Above, African elephant; below, its Asian relative.

gives much more strongly the impression of having been "molded in one," and therefore appears more aesthetically satisfying, more "adult." The Asian species is also rendered more childlike by virtue of its arched back, which makes its body bulkier, and by its smaller ears and shorter tusks, which in some cases, including the Sri Lankan elephant, are entirely lacking. These features distinguishing the Asian elephant from the African elephant are tendencies particularly characteristic of *human* morphology, in that the structuring of the body is not pushed to the limits of specialization; something is held in reserve, as it were. In humans this formative potential, which does not exhaust itself in fixed, structural specialization, is carried over into another sphere of activity—out of the natural, and into the cultural realm. Something similar would appear to be true of the Asian elephant. Its juvenile features, although much more rudimentary than in humans, are nevertheless much stronger than is usual in mammals. Their kinship to those of humans is indeed close enough to enable humans and elephants to enter into cultural partnership—the former, of course, providing the driving force.

Is Sri Lanka a paradise? No, only a pale reflection of it, a last glimmer. In fact, this account could almost be its obituary. What appears to be a possible model of how to treat nature in a caring and harmonious way in the future is actually only an echo from the past with no chance of standing up to the onslaught of the present. Only remnants remain of the original forests and wilderness, which daily become smaller. They have had to give way to tea, rubber, cinnamon, and clove plantations—the sterile, artificial landscapes produced by intensive agriculture. And the only reason there are still any elephants is that certain dry-forest areas have been declared nature reserves, where the gray giants can live undisturbed. Their free-roaming days are over, for they had been a thorn in the flesh of settlers who, understandably enough, were somewhat disinclined to tolerate them on their plantations. As human numbers increased, settlement pressed more and more into hitherto unpopulated areas, and humans and wild elephants cannot by any stretch of the imagination live together in the same place, sharing the same rice field! Driven into the forests, the pachyderms are forced into competing for food with the wild of semi-domesticated buffaloes, which are considerably more fertile, and multiply accordingly.

And not only in Sri Lanka is the elephant in a sorry plight—in fact on this island its circumstances are still comparatively good. Its range, once extending from Syria, indeed from Antioch in what is now southeast Turkey (Kumerloeve 1975), to China, has dwindled to a few, small, island-like refuges in South and Southeast Asia, and its numbers, of course, have dwindled accordingly. Today the Asian elephant stands under threat, although not the African elephant, which exists in much larger numbers.

The ultimate destruction of this paradise is inevitable. It may be, nevertheless, that not just its memory will remain to be passed on, but something else besides: the experience, the certainty that humans are capable of altering their sensibilities such that they can work together with nature in a more kindred and creative way. Perhaps the tropical island paradise is passing away because this precious attitude has been embedded in an unreflecting and passive mode of life. In the end it may have become mere habit and hollow tradition that can offer no resistance to the modern machinery of exploitation, which gathers its energies from quite another source: greed coupled with a highly refined intelligence geared solely toward the control of matter.

Perhaps only when this process has been pursued to its inexorable end, and the resulting total destruction is experienced as the insanity it is, will there be an awakening, a jolt of consciousness. The evening twilight into which the island paradise sinks betrays not a hint of this.

Yet for those with eyes to see it is clearly and prophetically expressed in the total artistic "synthesis" of nature and culture that is Polonnaruwa. The very fact that we experienced this place as a union of two separate spheres—nature and culture—shows our basic lack of true understanding of the spirit out of which the work arose. We merely interpreted it in terms of our modern split consciousness. The division of experience into an outer sense-world and an inner mental sphere of thoughts and feelings is something that we simply take for granted, and we have great difficulty accommodating the one to the other. Polonnaruwa is something wholly other, and its singularly harmonious perfection cannot really be expressed in words, for here this division does not figure. There is no inner or outer, but only the one undivided consciousness. Humans, nature, and the whole cosmos are one; Humans do not experience themselves as being alone, self-centered, "alienated," but as integrated into a whole. Something of this was surely brought home to us as we observed how nature attunes herself to human beings' moral and ethical attitude toward her. Does the one evoke the other, or do they interpenetrate in both time and space, even though our "inner-centered" consciousness is not aware of the fact? Are not the world within and the world without in truth one and the same?

Polonnaruwa is the purest and highest expression of Hindu-Buddhist culture. It expresses a state of oneness preceding our split Western consciousness: the paradisial state. In the Vedic creation story it is portrayed in the following words:

> Nor death nor immortality were then,
> Neither had night nor day come to be,
> In the beginning breathed windlessly
> The One, apart from which nought else was.

About this passage the orientalist Veltheim-Ostrau observed:

> From the point of view of human development this refers to the state of human beings before the Fall. The eating of the forbidden fruit is the source of the cleaving of existence into good and evil, day and night, man and woman, fortune and misfortune, birth and death, this world and the next. I distinguish these two fundamentally different states [of consciousness] in terms of, on the one hand, *undivided, open life,* the past and future paradisial state, and on the other, divided, closed life, by which our current existence is circumscribed. It is crucial for a deeper understanding of Nirvana . . . to realize that on the lips of Buddha it does not mean the Beyond, a word which in itself implies division, opposition, but rather a state transcending division, of perfectly reconciled oneness, achieved anew after passing through the divided state. (1954)

At this level there is also no difference between past and future; the one is contained in the other, indeed the one is the other. This means, of course, that however independent of time this state might be, its occurrence, no matter when, depends upon our actions, my actions. This, in turn, implies something that leads beyond Buddhist sensibility: the factor of development. What was once experienced by human beings in an unindividualized, participatory way—oneness with nature and the cosmos—must now, if it is to be experienced anew, be evoked by the individual through the combined effort of thought and will. The old state of things, as we saw, is doomed. It can be reborn in inverted form, but it is in the nature of this inversion, this meta-morphosis, that it will no longer come in the form of a revelation streaming into humankind from a surrounding spiritual atmosphere; rather it can be found anew and realized only inwardly, through the conscious efforts of free individuals. And it lies in the nature of this freedom, devoid as it must be of any hint of compulsion, that it can be wasted, neglected, and misused.

DEMETRIA: A Model for Restoring Life to Depleted Soil

The countries of tropical Asia have not yet entered this phase of con-sciousness—they are still too close to paradise. In the tropics of South America, however, things are very different. It is astonishing, but nonetheless understandable, that just in places where destruction has been greatest—in the urban wastelands of Brazil and areas worn out by intensive agriculture—more and more people, including some from rural regions, are on the lookout for new ways of halting and pre-venting it. For instance, the populations of Peruibe, Itanhaen, and Iguape in the south of Sao Paulo joined together with great courage, imagination, and ingenious publicity in a campaign to save the Serra da Jurea, a scenic jewel of the highest order, from the simultaneous construction of three nuclear power stations.

Most impressive, however, are the experiments seeking to restore life and fertility to worked-out, abandoned land. These are especially significant, because the first energetic attempts in the direction have already met with distinct success, pointing the way for others. Probably the most important of its kind is the biodynamic enterprise called Estancia Demetria. It lies in the interior, several hundred miles west of Sao Paulo, near the small town of Botucatu. Typically enough, the plan for this venture was city-born and conceived in the 1970s. The idea was to leave the overpopulated city, the city gradually suffocating in its own social problems, and to oppose the catastrophic flight from the land with a countermovement based on new initiatives. As a location for this initiative an area with particularly poor soil, destroyed by previous coffee cultivation, was purposely chosen. The young people who began it were not dependent on vigorous enthusiasm alone, but also knew what they were doing and had the guaranteed support of a backup organization; therefore they managed to pull through the first terrible years when no glimmer of hope showed and setback followed setback. Gradually, however, they did succeed in recreating fertile, arable land, and today the terrain is a blossoming paradise and a new one at that—youthful, assured of its future—not a doomed remnant. It has been fully self-supporting for some time now, and delivers its products as far as Sao Paulo.

This mixed style of farming, which ranges from growing fruits to cultivating medicinal and culinary herbs, as well as rice, with dairying thrown in for good measure, is very labor-intensive. This means that large numbers of people are required, and those who settle find themselves living in modest, but by no means poor, circumstances. Demetria is thus exemplary and forward-looking also from the social point of view. Ultimately the vast numbers of people who forsook the land (often under compulsion), only to end up vegetating in the indescribable conditions of the slums and shanty towns of the big cities, will have to return to the empty countryside—where else is their food to come from? The currently existing giant monoculture concerns—much of whose products are exported anyway—offer no sustainable prospects for life and work.

The constellation of personnel represents another good omen. It is Brazil in miniature, as it were, a motley collection of people from all over the world. In one family the mother is Norwegian, the father Brazilian-German: they speak English to each other and Portuguese to the children. There are, of course, some Brazilians, young agroscientists from the city as well as local peasants, but all are homeless in one way or another. And here they are beginning to establish a new home, and a community based not upon common descent, but upon common ideas and aims.

The health and dynamism of this new community shows in its capacity for change, its eagerness to rethink and redesign things, and in the multifarious alterations and extensions to the buildings. A workshop for agricultural research has recently been added and is in regular contact with Europe. It was also not long before the farm was teeming with boisterous children, who needed a school locally. What else could be done, then, but to found—in spite of all official obstacles—a small Waldorf school? A children's hostel, an offshoot of the Favela initiative at Monte Azul in Sao Paulo, has also settled in the neighborhood, and more and more people from the city are buying plots of land, both large and small, all around. In this way they become joint owners and place their land at the disposal of Demetria. This does not represent escape to a rural idyll (no longer attainable in any case)—to an island, certainly, but not the sort that seeks to isolate and screen itself off, rather one that radiates its influence to its surroundings and attracts in like measure. It is only a drop in an extremely wide ocean, when viewed in comparison with the vastness of the whole country with its no-less-enormous problems. Quantitative greatness, however, is of far less significance than qualitative. What matters is its effectiveness as a shining example for others.

Wandering through the rich variety of Demetria's fields and orchards after a long journey through bleak and largely uninhabited land, one feels as if suddenly relieved of a great, pressing burden; one begins to feel at ease, as if one had just returned to a long-familiar homeland. And one is surprised and delighted to find that here not only culture is present, but nature too has reinstated herself. Birdcalls sound on all sides, songs no longer heard in the surrounding landscape. Richly colored tropical butterflies (where could they have come from?) visit the many wildflowers and the welcome abundance of "weeds" thriving on the banks and beside the pathways. We even encounter a real beast of prey: arriving at night, we catch in our headlights a pampas cat surrounded by its young—just like a miniature leopard, but all the more elegant and graceful for its very slightness.

It's not too late after all, it seems. It really works. Nature is still prepared to cooperate and meet humankind halfway, if we approach her in the right way. This indeed we must do. Nothing happens of itself anymore.

Juvenilization in Evolution and Its Ecological Significance

THE EVOLUTION of living things is commonly thought of as a process in which the simple and primitive, by very gradual refinement, becomes complex and highly sophisticated. Although there is some justification for this view, it turns out to be entirely one-sided and thus fails to do justice to the full complexity of the process itself. A closer look reveals an intricate interplay of tendencies moving in different, indeed opposite, directions.

In the animal kingdom there are certainly developmental processes that fit the picture of progressive refinement. Early unspecialized forms, still, as it were, open on all sides, are seen to develop a more "limited" body plan (gestalt), in which certain structures are then taken to ever greater extremes of differentiation. Specific lines of development have their own particular "motifs," which are gradually perfected and thereby raised to their highest level of development. Among the animals examples of this mode of development abound. To take just a few: in the ungulates the early forms are unspecialized in bodily structure, omnivorous and therefore relatively unspecialized in their digestive systems (for example, the chevrotain, Tragulidae), while in the later forms we have the full variety of ruminants with their highly specialized gut. More extreme are the differences within the elephant family, where the line runs from the pig-sized and piglike Moeritherium of the early Tertiary period to the forms of the Pleistocene and the present day. Similar contrasts are found in the developmental history of the whales, while they are even more extreme among various orders of saurian reptiles of the Mesozoic. As a rule evolution does not simply follow one line; rather what we see are radiations of lines, each one representing adaptation to a particular way of life—in other words, one-sided structural modifications in connection with specific functions. For example, with its long legs the cheetah is committed to catching its prey on the run at high speed. The shorter-built leopard is not capable of such speed, but instead is a highly skilled stalker. According to their specializations, then, these two species inhabit different ecological niches—the former open plains, the latter thick undergrowth and rocky landscapes.

Many lines of extinct mammals and giant reptiles of the Mesozoic follow the same course. Beginning with simple, generalized forms with a great deal of adaptive potential, they soon lead to various structural extremes. Among the currently so fashionable saurians, for instance, we find flying forms, aquatic forms that seem to resemble fish or dolphins, giant herbivores, horned forms, giant two-legged predators, etc. Gigantism is indeed a common feature of the process, found not only among the "dinosaurs," but also many mammal groups: elephants, titanotheres (Fig. 1), rhinoceros, etc. It has even caused some birds to forsake the skies: moas, Diatryma, ostriches. These animals give the impression of having exhausted all possible avenues of structural change, some of them being pushed to truly awesome proportions—

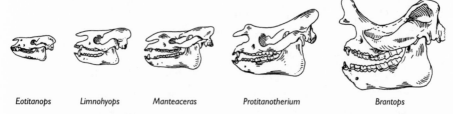

Eotitanops   Limnohyops   Manteaceras   Protitanotherium   Brantops

Fig. 1. Evolution of the brontotheres (*titanotheres*), a group closely related to the rhinoceros. It culminates in giant forms with wide, shovel-like tusks. They first appeared in the Eocene and had already died out by Oligocene times. The phylogenetic sequence probably gives a fair approximation of their ontogenetic development, with the smaller and simpler early species corresponding to the juvenile stages of the larger and much more elaborate later species.

as, for instance, in the tusks of the elephant family. These are all instances of *peramorphosis*, evolutionary ageing. In many cases the next step is extinction.

The process of specialization and the specific ecological relationship that goes with it lead inevitably to an ever increasing limitation of adaptive capacity, and ultimately into an evolutionary dead end. The evolution of the horse is one example of this among many. What begins as a browser changes to a grazer with limbs attuned to life on the open plains, and in conjunction there is a transition from the five-toed to the three-toed, and finally to the one-toed hoof of the present. Any further development here is somewhat inconceivable.

The polar opposite to the widespread developmental tendency just described is one in which evolution is not a gradual ageing, but a process in which successive "juvenilization" takes place. This counter-tendency, also termed *neoteny* or *paedomorphosis*, has become well known because of the role it plays in human evolution. In this connection it has been so often described and investigated that here we need only present a brief sketch.

In Schindewolf's well-known diagram (Fig. 2) the skulls of our immediate ancestors are arranged in chronological order. This brings to light a gradual reduction of growth in the jaw region, which becomes increasingly less prominent in relation to the cranium. Each

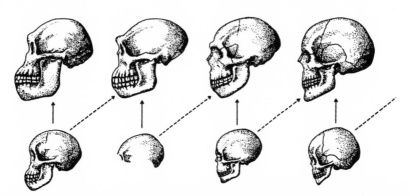

Fig. 2. The cumulative "juvenilization" of the Hominidae, shown by comparing the juvenile (below) and adult phases (above) of, from left to right, *Australopithecus, H. pithecanthropus, H. neanderthalensis,* and *H. sapiens* (right). Each succeeding adult form ceases growth more or less at the juvenile phase of its antecedent. (After Schindewolf 1972.)

successive (later), adult skull adopts a form more or less corresponding to the fetal stage of the previous one. The jawbones become increasingly youthful, paedomorphic. The same goes for the hands. Their
undifferentiated fivefold form is primeval, primitive in comparison
with the specialized forelimbs of most vertebrates, and thus bears a
close resemblance to the basic plan of the limbs of the ancient
tetrapods.

It should be pointed out that here only certain organ systems are
affected—by no means is the structural organization of the whole
organism subject to paedomorphosis. On the contrary, the brain above
all, but also the feet are just as highly differentiated in humans as the
jaw and the hands are youthfully general in form. Such a "mosaic
structure," combining primeval and highly developed features in one
organism, is in no way confined to human beings, but is a widespread
phenomenon. To name just one example, it is found in birds, which
in some respects are the most highly developed of all vertebrates. They
have the most perfect respiratory system, the sense of sight with the
highest acuity, and, in their feathers, the most complex skin structure,
while at the same time their feet are only minimally supplied with
blood and warmth and are covered with the same horny substance as
the feet of reptiles. Nevertheless, in almost no other living thing is the
polarity between paedomorphic and peramorphic organ systems as
great as in the human being.

Having a mosaic structure does not mean, however, that the organism concerned is some kind of conglomerate of more or less independently evolving organ systems. Rather we could speak in terms of
different streams of development, and of each particular species being
a focal point where they meet in a unique way. The peramorphic stream
we began with may be designated as determined by the past. A developmental line, once started, is preserved and consistently pursued to
its endpoint. We see this, for instance, in the deer family: the two small
prongs set in the forehead of the early forms are gradually increased
in size and complexity until they attain the tremendous weight and
magnificent scale of the antlers of the giant deer ("Irish elk," *Megalo-
ceros*). Throughout this group of animals the form of the antlers takes

every possible turn imaginable, *but nothing new emerges;* it is all ultimately variations on a single theme. Something similar occurs in the proboscidians (elephants, mastodons, mammoths, and their kin), except that here the variations are played upon the incisors (Fig. 3).

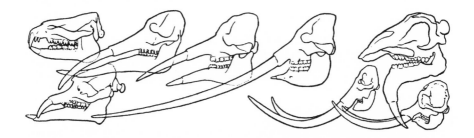

Fig. 3. Variations of the elephant motif. Apart from the African elephant (below right) and the Asian elephant (not shown), all species in this group died out in the course of the Tertiary and the Pleistocene periods. Moeritherium (above left), from the early Tertiary, is the oldest, most primitive representative of this family (scale varies between illustrations).

The process of paedomorphosis would appear to act in polar opposition to this. Here functional and structural development is held back; the forms of organs remain at an indeterminate stage and therefore *open to innovation.* It is significant, as Wolfgang Schad has pointed out in a recent study, that *paedomorphic forms are found particularly at the crucial moments when major macroevolutionary innovations occur.* They seem to function as a source of new evolutionary impulses at all the major transitions: from fish to amphibian (crossopterygians), from reptile to bird (Archaeopteryx), from reptile to mammal (therapsids).

PARALLELS IN THE PLANT KINGDOM

The interaction of contrary developmental tendencies not only applies to the human and animal realms, but also plays a key role in the development of the higher plants. From this flow consequences of both major evolutionary and ecological significance. Strange as it might at first sound, the contrary tendencies create the conditions whereby the currents in the development of human being and plant not only impinge upon, but interpenetrate and mutually enhance each other. This will be described in what follows.

The forerunners of the ferns, archaic land plants from the Silurian and early Devonian periods, were still mostly amphibious (Fig. 4) and showed little in the way of structural differentiation, certainly no clearly defined organs such as leaf and stem. Tendencies to leaflike structures do occur here and there in the form of small, flattened scale-like outgrowths reminiscent of the microphylls of mosses and liver-

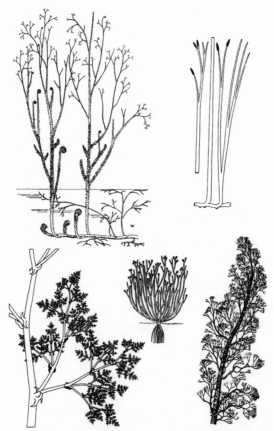

Fig. 4. Archaic land plants, early relatives of fernlike forms with a simple dichotomous growth pattern inherited from the algal phase. Above left, *Asteroxylon elberfeldense* (after Mägdefrau 1968); next to it, *Rhynia*; in the middle *Hicklingia* (after Zimmermann 1969); all Lower Devonian species. *Pseudosporochnus* of the Middle Devonian (below right, after Zimmermann 1969) and *Stauropteris* of the Lower Carboniferous (below left, after Mägdefrau 1968) still follow a simple growth pattern of dichotomous branching in all directions—in *Stauropteris* what looks like a leaf is really bunches of smaller and smaller side shoots.

worts, and are perhaps their forerunners. They are thus not homologous to the vegetative leaves of the ferns and flowering plants, but may be regarded as functionally analogous to them, as are the tiny leaf-scales of mosses (genuine leaves only arise later with increasing "division of labor" within the plant organism). The psilophytes, early forerunners of the ferns and club mosses (Lycopodiaceae) and the horsetails (Equisetaceae), had a bushy growth pattern brought about by simple (dichotomous) forking of the shoots in all directions. During the course of the Devonian they gradually became more centralized: a main axis began to dominate through monopodial growth and consequent reduction in the growth of side shoots (Fig. 4). The side shoots could, if anything, be said to fulfill the function of leaves, even though there could be no question of genuine leaf structures at this stage: the shoots still fork in all directions (Fig. 5). In other words, there is still no sign of any kind of organ in the form of a closed surface. The integration of the parts into a composite form first occurs in the late Devonian. Plants like *Archaeopteris* (Fig. 5) form side shoots with their branchings all arranged in one plane, thus resembling genuine fern fronds. With this, succeeding the monopodial growth by which a central axis was formed, the second "basic process" (Zimmermann 1965, 1969) was complete, namely, planation, or flattening. Venation, however, remained primitive, being forked or fan-shaped; pinnate or meshed venation was still to come.

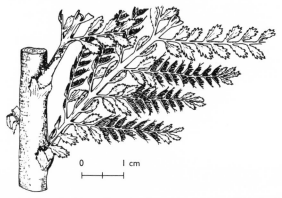

Fig. 5. *Archaeopteris* from the Upper Devonian (from Zimmermann 1969, after Beck), in which we see the gradual emergence of the flat, dorsi-ventral fern frond with fertile spores on some of its pinnae.

What is merely hinted at here attains its full expression from Carboniferous times onward in the typical fern frond. No longer do we have just an aggregate made up of arbitrarily repeated side shoots, but a unified form to which the parts are subordinated. The individual shoots have thus been integrated into a composite whole: "The more imperfect a living being is, the more its parts resemble both each other and the whole. The more perfect a living being is, the more unlike each other are its parts. In the former case the whole is more or less identical to the parts, in the latter parts and whole are dissimilar. The more similar the parts are, the less subordinate they are one to another. Subordination among the parts is an index of perfection" (Goethe 1817). In passing it may be mentioned that this polarity between differentiation and centralization will later repeat itself on a higher level with the "subordination" of arrays of multiple blossoms to the formative regulation of a single "master blossom" (Pseudanthium) in the Compositae.

Fig. 6. From shooting to subdividing: fronds of recent ferns. The repetition of the whole in the part— of the whole frond in primary and secondary pinnae—is still detectable, and, exactly like the frond as a whole, each pinna has its growth center at the tip (and not at the base, as do the leaves of flowering plants); shoot and leaf have not yet been "uncoupled," as in the higher plants. At the same time the new principle of the "subordination of parts" is already apparent to a greater or lesser degree: the single pinna is no longer simply a repetitive element, but is integrated into the total form in obedience to some higher principle of order. The species here shown are hard shield fern *Polystichum aculeatum* (top), mountain bladder fern *Cystopteris montana* (middle), canary lady fern *Diplazium caudatum* (bottom).

The next step in development has already been prefigured in the phenomenon of centralization: the branched shoots begin to lose their spatial separateness and are gradually integrated into the total form: the unified leaf surface begins to take shape (the third of Zimmerman's "basic processes"). This tendency increases significantly during the Carboniferous period (Fig. 7), and what appears in the whole repeats itself in the part. Venation is at first centrifugally fanlike (Goethe would have called this the "expansion" phase), and gradually becomes subject to centripetal influences ("contraction"), which lead initially to the formation of a central axis, although the "centrifugal" tendencies persist at the periphery (pinnate venation). Finally this process of

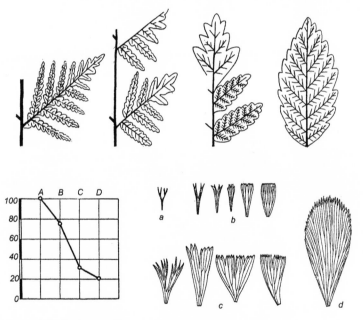

Fig. 7. From subdividing to spreading.

Upper row: frond development in the fern family of the *Emplectopteridae* from successive layers of the Permian period.

Center: curtailment of free-growing shoots and transition to the closed leaf blade in the fern group Sphenophylum as shown through various geological periods. Left: Horizontal axis in the graph: A. Upper Devonian; B. Lower Carboniferous; C. Upper Carboniferous; D. Permian. Vertical axis: percentages of known species with free-growing shoots. Right: a. *Sphenophyllum tenerrimum* (Lower to early Upper Carboniferous); b. *Sph. cuneifolium* (middle Upper Carboniferous); c. *Sph. maius* (middle Upper Carboniferous); d. *Sph. thoni* (Upper Carboniferous to Permian).

Bottom row: netted leaves of varying intricacy. From left: *Glossopteris*, a pteridosperm (seed fern) from the Permian. The others are all living forms: gray willow *Salix cinerea*, bog willow *S. myrtilloides*, barberry *Berberis vulgaris*, and wild ginger *Asarum europeum*.

contraction also takes hold of the periphery and the anastomoses (reticulated venation) then weave the endings of the veins into a closed, unified, functional whole. It may well be that the correspondence this bears to development in the growth forms of woody plants is more than merely external analogy. At first we have bushy growth with no central axis; then the simple, regularly branched monopodial growth form of many conifers (monkey puzzle tree [*Araucaria*], young spruce and firs); and finally the sympodial, broad-leafed tree with its integrated branch structure.

Other plant families reveal further analogies of this kind, insofar as the developmental steps associated with them were underway by the end of the Paleozoic—for example, both types of ginkgo (Fig. 8). The leaf finally attains its full form with the dawning of the angiospermous flowering plants.

Three main steps have been identified in the evolution of the leaf:

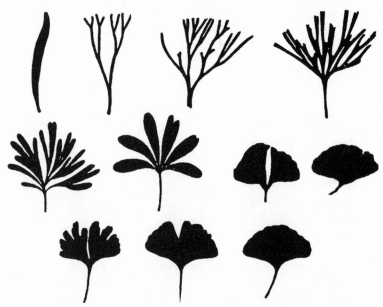

Fig. 8. Morphological changes in the phylogenesis (upper two rows) and ontogenesis (lower row) of the ginkgo leaf. Above, from left to right: *Glossophylum florini* (Upper Triassic), *Sphenobaieria furcata* (Upper Triassic), *Sphenobaieria digitata* (Upper Permian), *Baiera muensteriana* (Upper Permian/Lower Jurassic). Middle row, left to right: *Baiera brauniana* (Lower Cretaceous), *Ginkgoites pluripartitus* (Lower Cretaceous), two leaves of *Ginkgo adiantoides* (Pliocene). Below: leaf ontogenesis in *Ginkgo biloba*, a living species. From left: leaf from a young tree 1 meter tall; leaves from a long and from a short shoot of a mature tree. (After Mägdefrau 1968.)

1) The formation of undifferentiated, scalelike structures, or repeatedly branched, identical shoots (phyllomae). This formative process could be called "shooting," in keeping with the terms used by Bockemühl (1966, 1995), to which we will return later.

2) The emergence of structural regulation, observable from the fact that formerly separate shoots become increasingly subordinated to, and indeed incorporated into, the formation of a central axis. A composite whole is now discernible, and the formative process involved has been called "subdividing."[1]

3) The parts lose their last remnants of independence and are incorporated into the very body of the overall form. The same is repeated in the pattern of venation—centralization is no longer focused only on the midrib, but takes hold of the branchings, joining them together. The leaf-blade is the outcome, and the formative process behind this development may be called "spreading."

JUVENILIZATION IN THE ANGIOSPERMOUS FLOWERING PLANTS

The biogenetic law[2] is usually illustrated by examples from the ontogeny and phylogeny of animals. Its application, however, can equally well be extended to the plant kingdom, for there are indeed many plants in which ontogenesis reveals a characteristic recapitulation of the changes in form that occurred in phylogenesis. Two of these are shown in Figure 8. This straightforward recapitulation, however,

---

1. To the best of my knowledge this is the first time the process has been called "subdividing." It is a rendering of the German word *Gliedern,* which has the simultaneous meaning of "dividing" and "integrating" or "ordering." Elsewhere it has been translated as simply "dividing" or even as "indentation," but here "subdividing" seems to catch the idea fairly well. — Trans.

2. This refers to the pattern of regularity first made widely known by Ernst Haeckel at the end of the last century, by which the embryonic development of the individual organism (ontogeny) is seen to recapitulate its evolutionary history (phylogeny). The recapitulation, it must be said, does not always display a one-to-one correspondence to phylogeny. —Trans.

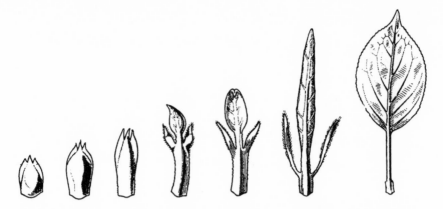

Fig. 9. From shooting, via subdividing, to spreading and elongation. Leaf sequence from bud-scale to mature leaf in apple (*Malus domesticus*). According to true scale the preceding stages should be much smaller than the finished leaf. At the base of its stalk are the two scars left by the stipules, which are present in the three preceding phases, becoming ever more subdivided. In discarding them the subdividing stage is superseded. (After Troll 1969, amended.)

holds only for woody plants, not for herbaceous forms. In the latter it appears only at a lower level of organization, that of organogenesis, in which the individual leaf can be seen to go through the three phylogenetic phases of "shooting," "subdividing," and "spreading."

To introduce a further qualification into the picture, in the apple, as in some other woody angiosperms, this three-step recapitulation appears only on the long vegetative shoots (Fig. 9); on the short, flower-bearing ones it behaves quite differently. Their leaves, like all leaf-forms close to the blossom, are reduced, held in check, as it were. In the apple the morphological transition occurs abruptly, in one jump, from normal vegetative leaves beneath each group of blossoms to very slender, delicate bracts at the base of each individual flower stalk. The transition is smoother and more gradual in the case of the wild rose, whose stalks culminate in single blossoms, or in the bushier peony (Fig. 10). Toward the calyx a gradual inhibition of the leaf blade occurs in conjunction with an expansion of the leaf base. It is exactly the same morphic movement[3] that occurs in the transition from leaf primordium to normal leaf, only in the opposite direction. The con-

---

3. This is a rendering of the German word *Bildebewegung*. Where a leaf is succeeded by another one of a different form, such a movement has taken place. It is only by an inner effort of imagination that such a transition can be seen as a movement. The cultivation of this "exact sensorial imagination" is the cornerstone of Goethe's approach to science.—Trans.

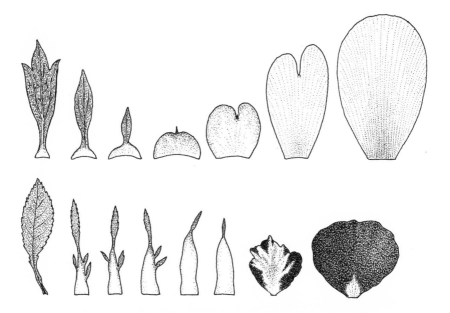

Fig 10. Morphic countermovements in the sequence of leaves from bud-scale to vegetative leaf (left section of both rows), and from vegetative leaf via the calyx to the petal (right section). Above: tree peony *Paeonia moutan*. Below: garden rose *Rosa sp*. In both cases the intervening sequence of mature leaves has been omitted

clusion to be drawn from this is that in the transition from normal leaves to bracts and sepals each succeeding leaf is arrested at an ever younger developmental stage.

This metamorphosis of the vegetative leaves appears in only a few woody plants in the pronounced way it does in the peony, and although it appears only somewhat cryptically in most of them, on nonflowering branches it remains true to the phylogenetic sequence. In herbaceous plants, on the other hand, it finds complete, and often highly differentiated, expression. Indeed in many of them it is the dominant morphological feature.

That leaf metamorphosis is a process of juvenilization was first shown by Bockemühl (1966). The growth phases of the individual leaf represent a truncated recapitulation of phylogenesis. It begins in the bud with "shooting," then comes a "subdividing" phase, and then comes "spreading" (combined with "elongation"), which results in a full (more or less), rounded leaf blade (Fig. 11). As each individual leaf emerges in turn, it ceases growth at a younger phase of develop-

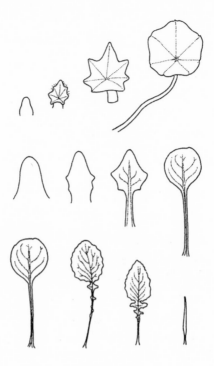

Fig. 11. In the upper and middle rows, from left to right: growth stages of a lower leaf from bud onward in *Nasturtium tropaeoleum* (above) and nipplewort *Lapsana communis* (middle). The bottom row shows four mature leaves of nipplewort, beginning with the finished lower leaf from the middle row and ending with an upper leaf close to the flower. (After Bockemühl 1966, altered.)

ment. While the lowest leaves reach the spreading stage, the middle ones halt at subdividing, and the highest ones get no further than shooting.

Thus it is *organogenesis* that recapitulates phylogenesis. Where this occurs fully in the lower leaves it runs through the complete ageing process. In the *ontogenesis* of the whole plant, however, the astonishing fact is that this sequence is progressively suppressed in favor of a morphic countermovement: *in the morphology of its leaves the herbaceous plant begins at the phylogenetically most advanced stage, each successive one representing a step "backward" to ever more juvenile phases.* This progressive *juvenilization*, both in the ontogenetic and the phylogenetic sense, signifies a reversal of the biogenetic law, a phenomenon first remarked upon by Alexander Braun (1851), and corroborated, in all likelihood independently, by Gerbert Grohmann (1931). The path just taken by this exposition from woody to herbaceous plants corresponds to that taken in evolution by the flowering plants. In evolutionary terms tree and shrub are older growth forms than the herbaceous plant. Currently the most active investigator in this unjustly neglected field of evolutionary research is the Armenian A. Takhtajan. His findings are so significant that what follows will largely be based on them.

Among the angiosperms the more primitive taxonomic groups are almost exclusively represented by woody forms (Magnoliales, Annonales, Winterales, Laurales), while in the more advanced orders herbaceous forms dominate. The young shoots of woody forms have anatomical features similar to those of herbaceous forms. Studies comparing closely related woody and herbaceous species found that in the latter secondary thickening (i.e., the laying down of woody tissue) was reduced such that their stems were essentially equivalent to the first growth rings of their older woody relations. The evolutionary step from woody to herbaceous forms is characterized by the reduction and ultimate cessation of cambium activity.

> The herb may be considered as a fixed juvenile form of the tree. . . .
> The herbaceous plants originated from the trees through neoteny.
> This process of neotenic transformation may be traced particularly
> well in those genera having both herbaceous and woody representa-
> tives. One of the best examples is furnished by the genus *Paeonia*.
> (Takhtajan 1991)

Naturally changes sometimes occur in the opposite direction, although the resulting forms are fundamentally different from their woody forebears: "Palms and bamboos are as different from primitive pre-angiospermous shrubs and trees as whales and seals are from fishes" (Stebbins quoted in Takhtajan 1991).

Of course, this evolutionary step is not the end of the story, for the herbaceous growth form, once established, continues to be subject to the process of neoteny. Again we follow Takhtajan on this point:

> In comparison to most dicotyledons the typical monocotyledons
> (including all primitive forms) appear to have undergone a degree of

"infantilization" at the vegetative level. They display certain simpli-
fications which give the impression of a prematurely concluded ontogenesis. Cambium activity is here displaced into axial organs, and
growth in the main root soon comes to a stop. The leaves either
remain uniform or show only an indistinct division into stalk and
blade, and have a pattern of venation like that found in the incom-
pletely developed leaf forms of dicotyledons (bracts, bud-scales,
sepals, etc.). These facts have led me to suppose that neoteny has
played a decisive role in the emergence of the monocotyledons....
Postulating such a neotenic origin seems to me to be the key to understanding many of their morphological peculiarities. (Takhtajan 1991)

The development of the monocotyledons is thus a further instance
of an evolutionary jump occurring by juvenilization. Here the progressive holding back of the evolutionary ageing process was pushed
further in that the very step that led from the woody to the herbaceous dicotyledons—this "regression" from a "fully developed" to a
"more infantile" form—was in turn removed from the pattern of recapitulation.

THE ANTICIPATORY NATURE OF NEOTENY

Juvenilization means that the process of specialization is held in check,
thereby creating a potential for new development. It is the tendency in
evolution that counters the linear prolongation of an established line
of development. As such it opens the door to novelty. "Neotenous rejuvenation" increases evolutionary plasticity and opens up new evolutionary pathways (Takhtajan 1991). "It is this possibility of escaping
from the blind alleys of specialization into a new period of plasticity
and adaptive radiation" (J. Huxley 1954).

There is a crucial distinction to be made here: while the process of
specialization represents the continuation of a trend set by *past* conditions, juvenilization opens the way to formative influences from the
future. These are not the causal consequences of neoteny, nor during
this latter process will any physical antecedents or rudiments have

been laid down. The point about the openness created by neoteny is that it is *indeterminate*, and this very indeterminacy leads one to surmise that the future form, though not prefigured physically, that is, genetically, is present as an ideal dimension within the process of juvenilization, and as an active principle creates the conditions for its own realization (Suchantke 1995).

In plants the source of novelty lies in the *development of the blossom* (in the broadest sense). This relationship is made extremely clear by the fact that in herbaceous perennials the juvenilization process is only observable on flower-bearing shoots. Those which have not flowered carry leaves that are all identical in form and have reached the "oldest" stage of development, that of spreading (Fig. 12). The blossom itself also bears the marks of neoteny, as direct observation of a plant in flower will reveal. All the visible organs of a plant begin their growth at the bud stage, but the closer the plant comes in its growth to the formation of blossom and fruit, the less its organs get beyond the bud stage. Whereas the lowest, first-formed leaves (in evolutionary terms

Fig. 12. In herbaceous plants metamorphosis only occurs in flowering shoots. Young, nonflowering shoots tend to put out leaves all the same shape as the lowest leaves on the flowering shoots. They thus display the most elaborate and phylogenetically recent leaf-form. Left: leaf-sequence of a flowering shoot (above) and a nonflowering rosette (below) of the scabious species *Scabiosa lucida*. Right: Asian crowfoot *Ranunculus asiaticus*, a young nonflowering, and a flowering sample.

the most highly developed) unfold fully, often varying strongly from each other in shape, this expansive gesture is more and more curtailed toward the blossom. The blossom, in turn, does not expand like the leaves; it remains bowl-like, its parts bunched around a central point, an arrangement arising from the fact that the stem has ceased to grow. Its growth has come to rest, as it were, at the early budding stage in which all organs are still in process of formation and there has not yet been any axial growth. With the development of composite blossoms in which the petals are merged, this tendency toward a closed form is intensified, with the fruit finally appearing as a "leftover bud," having kept to the very earliest growth stage under its covering of (neotenous!) carpels.

The petals follow the same pattern of neoteny, whether derived, as in most cases they are, from the growth of stamens, or from that of sepals (as in the peony). The same applies, albeit even more strongly, to the carpels.

This process of neoteny, thus expressed by both petals and carpels, creates the conditions for two completely new morphological motifs that came with the emergence of the angiosperms:

1) In place of the serial organization hitherto displayed in the plant's overall structure, we find in the blossom a composite whole in which several organs have been combined in geometric regularity and more or less constant numerical relationships. All this has been further enhanced by color and fragrance. With these new qualities the plant has attained an "image," an *expressive* form that implies an "other" with perceptual faculties capable of being *impressed* by it. Here the plant touches the world of sentience.

2) The process of double fertilization develops, which is unique to the angiosperms. This entails the female gametophyte reverting, through neoteny, to an archaic phase wherein it consists of a formless, multinucleate plasmodial embryo sack. Into this penetrates a pollen filament containing two (haploid) sperm nuclei. One of these merges with one of the (haploid) nuclei of the female embryo sack, thus completing "normal" fertilization. The

other one unites with an egg "cell" consisting of two merged embryo-sack cells. The latter is thus diploid and through this "second" type of fertilization then becomes triploid.

One of the fertilizations leads to the formation of the actual embryo, while the other is concerned with the laying down of the endoderm (nutritive tissue). This is a true innovation generated by the angiosperms; in other words, it is not a phenomenon that has slowly developed out of primitive antecedents, not simply the continuation of an ancient line. It was made possible through the process of juvenilization that began in the leaves and was intensified in the blossom, where it ultimately took hold of seed formation. Here we see the morphic potential of the plant being held back from full expression at one level and directed instead into the shaping of entirely new structures.

Juvenilization, neoteny or paedomorphosis, quite clearly appears to be a higher formative principle operating on the same level as the ontogenetic recapitulation of phylogeny. The question immediately arises, What are the implications of the parallel occurrence of this evolutionary gesture in human beings (the vertebrates) and in plants? Is it even possible to make a valid comparison between tendencies toward juvenilization in human beings and in plants? To attempt to draw parallels between the actual formations of particular organs would obviously be pointless. Nevertheless, the basic gesture of juvenilization followed by morphological and functional innovation (neomorphosis) remains clearly observable in both. Moreover, the same seems to be true of the biogenetic law, for its expression is not confined to strictly homological structures, but extends beyond the level of genetic relationship.

In the case of human beings, development has stopped short of the full functional specialization of certain physical organs, thus giving them the potential for something new. The spare formative potential is then at the disposal of this "something new," whatever it might be. Thus, for instance, the paedomorphic human jaw set up the potential for the acquisition of language (Kipp 1955, 1985), and the hands, in "remaining unspecialized," became multifunctional—and the new functions are consciously learned rather than genetically fixed, as is largely the case in animals. Thus the stage was set for the development of a "second realm of nature," nature raised into the sphere of human culture.

The higher nature of the evolutionary principle involved here is especially underlined by the striking way in which the evolution of human beings and that of flowering plants are interrelated—a sort of coevolution on a higher level. There would seem to be a single principle at work in the development of both. With this we enter the ecological sphere.

ECOLOGICAL ASPECTS OF JUVENILIZATION

Two evolutionary achievements of the angiosperms are of decisive significance both for the overall economy of nature and for human culture: the herbaceous growth form and double fertilization.

The latter enables flowering plants to build up and store large quantities of nutrients in their seeds. This combined with the herbaceous form of growth means that seeds form a relatively high proportion of the total phytomass, very much higher than in woody plants. This applies especially to the grasses, which are extremely frugal regarding their vegetative organs. Trees present exactly the opposite picture, the greater part of their biomass, wood, playing no part in living processes. The transition to the herbaceous growth form also meant an acceleration in the breeding rate, and consequently an increase in species diversity due to the quicker appearance of new forms. This in turn provided the basis for a great upsurge in higher forms of animal life, most of which were mammals: "The development of the herbaceous flowering plants was tremendously important for the evolution of animals, especially herbivorous mammals and ground-dwelling birds" (Takhtajan 1991). From the beginning of the Eocene onward a buildup of grass pollen is observable in sediments, and then from the middle of the Miocene there is clear evidence of the steadily increasing spread of open grasslands. Evidently at this time large regions were being transformed into savanna-type landscapes (Daghlian 1982; Thomas and Spicer 1987).

This was the formation of those "superorganisms," the great grassland ecosystems (biomes) in which tree growth is inhibited by periodic drought. In temperate latitudes they are familiar to us as the steppes and prairies, but it is in their tropical form that the reciprocal relationship between plant and animal evolution appears in its full significance. In the African savanna the grass shoots up in the rainy season and is cropped by the herds of large grazing animals. The dung they leave is swiftly dispatched underground by scarab and dung beetles, where the next drought's fires (caused by lightning) cannot destroy it. The heat of the fires is never very intense, for the successive herds leave nothing to fuel it—the last ones through, the gazelles, even eat the stubble. As a result, the heat does not penetrate even a few centimeters into the soil, and the grass roots remain intact.

The ecological relationship between grasses and grazers is thus a very close one, further enhanced by the fact that the same large herbivores prevent the land from being overrun by thornbushes.

This example illustrates in stark terms just how essential the ecological dimension is to any understanding of evolutionary processes. After all, living things do not develop in isolation, but always in mutual interaction with other organisms and with the climatic and soil conditions of their surroundings. Evolution is always coevolution, and this holds not only at the level of the individual organism, but also, as the above example shows, at the level of the ecosystem as a whole, the "ecological organism," in which particular animal populations and plant communities function as specific organs.

It might appear at first glance that we now have a contradiction: whereas before the talk was of novelty, of evolutionary tendencies "prefigured" by juvenilization, now it would seem that the ecological dimension is all-important. On closer inspection these appear as two sides of the same coin, as is shown, for instance, by the interplay between hominid and angiosperm evolution. Here, through the agency of a higher formative principle, species attain their characteristic expression and mutually enhance each other's development. This is comparable to the formative processes that take place in the growth of an organism. With increasing maturity its parts—organs, cells—come to resemble each other less and less and take on forms specific to their various functions. At the same time their interrelationships and mutual dependencies reveal ever more clearly the regulatory, formative influence of the whole at a higher level of organization.

Of course, "ecological organisms" (ecosystems) are not identical to single organisms, such as individual plants or animals. They are organisms of a higher order with their own specific features; for instance, they have a greater degree of openness, and are more vulnerable to change, typically at the hand of humans. Nevertheless, they also display the complementary aspects just mentioned, and this, more than anything, demonstrates how blurred the distinction is between the single organism and the ecological organism. A comparable pair of complementary features have been identified as "reproduction" and "nutritive multiplication" (Steiner 1924). On the one hand, the individual organism, indeed the population, is geared toward the preservation of the species through reproduction, while on the other hand a contribution is made to the growth and persistence of other living things by surplus production of biomass, either in terms of numbers or body size. (A similar interplay is also found between the regulatory activities of consumer organisms and of the decomposers, who break down the detritus and thereby keep the mineral cycles turning.) Organisms, it would seem—in a figurative sense, of course—have an "egoistic" and an "altruistic" side to their natures. In passing it may be mentioned that the same principle applies to human beings—as individuals—within society.

The scene of the most recent phase of evolution was not the forest, the domain of the woody plants, but the landscapes where neoteny holds sway in the form of herbaceous plants, and among them the most "juvenile" forms: the monocotyledonous grasses. Even today awareness of this fact tends to be lacking. Attention is more likely to be focused on the tropical rainforests, under threat as they are, and with their vast array of species in obvious and urgent need of protection. There, however, the scene is not dominated by the higher species—herbaceous plants and warm-blooded animals—but above all by the insects, which have developed a truly luxuriant abundance and diversity. For the higher, warm-blooded animals conditions there are very unfavorable, while herbaceous plants are either excluded or eke out a marginal existence in the forest canopy. Because of a shortage of protein in the forest food chains (Reichholf 1990), the former tend to be small (duiker, royal antelope, dwarf forest elephant, dwarf forest buffalo, etc.) and never form large herds. Larger warm-blooded forest-dwellers roam solitary, as, for instance, the okapi (cf. chapter 2). The fact that the greater part of available fodder is not found on the dim forest floor, but in the canopy, is a further handicap for ungulates; most browsers are forced into the treetops (sloths, howler monkeys, langurs, etc.). What applies to the mammals is of even more relevance to humans. Genuine forest peoples like the Ituri Pygmies, or the numerous groups of rainforest Indians in South America, have remained in their state of primal integration with the forest, largely unchanged since time immemorial.

The scene of major cultural evolution, however, has always been the open, "juvenile" landscapes. And that is not all. Animal-rich grassland is the place of humankind's first emergence: the African savanna, where there is no lack of tree cover, but so thinned and well spaced that a complete carpet of grass has been able to take hold. Of all natural landscapes it comes closest to perfection. No large-scale ecological organism is more richly endowed with living things or more diverse in its way than this one. Its flora displays a wide variety of growth forms without the tree—which is "older" in the evolutionary sense and inhibits the development of more "juvenile" forms—being dominant. This richly varied vegetation provides a context in which the animal world can flourish, the mammals and birds, as well as the insects. And this powerhouse of evolutionary processes is suited to human life in a way unlike any other landscape system. It is ultimately responsible for humankind's comparative unattachment, endowing us—unlike the animals, always bound by instincts and specializations to a particular niche—with the potential for cultural diversity and the practically unlimited development of new abilities. These, however,

Fig. 13. In the African savanna the sparse distribution of the trees (*Acacia spp.*) permits the development of a very dense grass flora.

do not arise entirely of themselves, but in relation to, and indeed through the struggle with, the environment. This in turn is repeated in the ontogeny of every human being: the richer and more stimulating a child's environment is, the greater is its chance of transforming its birthright of universality and openness into a rich range of abilities.

There is, in fact, a broad consensus about the significance of the savanna for the cultural development of humankind, exemplified in statements such as "Our ideal landscapes are those which resemble the savanna" (Eibl-Eibesfeldt 1984), or "The park landscapes we have created . . . may be interpreted as an expression of memory reaching back into the long history of nature" (Gerken 1995). It would seem that human beings carry within them an image of "their" landscape, and re-create this inborn image of their ancient homeland wherever they go (Suchantke 1986, 1993).

Thus it is not only the birth of humanity that takes place in the open "juvenile" landscapes. They also provide the background upon which the crucial steps in the development of human culture are taken. On the wide, grassy savannas of the Near East, the step was taken that made humanity an active participator in the shaping of evolution (Suchantke 1986, 1996). The Fertile Crescent is certainly the most productive of those cultural focal points where plants and animals underwent domestication and selective breeding, wheat and cattle being the prime products of the process in this particular area (Zohary and Hopf 1993). Grass, this most juvenilized of all plants, which develops the "altruistic" principle of *nutritive multiplication* to its highest pitch, is the basis of practically all cultures. In terms of quality and quantity, the basic foods of all major cultures have been provided by three grass species: wheat, rice, and maize. And it is no exaggeration to say that without grass there would have been no human cultural development.[4]

A third phase began on the western margin of the ancient world, with the spread of arable agriculture into Europe from around 6000 B.C. onward (Howell 1989; Lüning and Stehli 1989; Zohary and Hopf 1993). This has continued right into the present and its history is a story of forest clearing. Where formerly closed, deciduous forest stood, offering no space for humans and their crops, the historic agricultural landscape gradually emerged. Today there are only remnants to show us what it was like. In its physiognomy and ecology it bore a very close resemblance to the savanna landscapes in which the first phases of cultural development took place, so close, indeed, that the first botanists to study East Africa at the turn of the last century named one form of savanna they found there "orchard steppe" (Engler 1910). The resemblance is created by open meadows interspersed with single trees and coppices—an artifact of human culture in Europe, in Africa a natural landscape. In the rainy temperate latitudes the forest would

---

4. Of course, there are apparent exceptions to this, for instance the Andean cultures, which have relied heavily upon tuberous plants such as the potato. But even there cereal-like plants are still cultivated for their high yield of grain, e.g. quinua, a "pseudocereal" of the goosefoot family (Chenopodiaceae). No matter what form it takes, agriculture is unthinkable without the use of herbaceous crops of some sort. Even bananas fall into this category, and palms are not originally of woody stock; rather they are monocotyledons which have reverted to the arboreal growth form, while preserving the abundant fruit and seed formation so characteristic of herbaceous plants.

soon return in the absence of human intervention and completely displace the flowery meadows.

This newly created landscape, however, may be described as artificial only in a very superficial sense. In reality it is just as much a product of natural process as it is an artifact, for in shaping it humans merely intensify a development that had already been begun by nature. There can be no question, then, of humans acting contrary to nature—certainly not while they are involved in developing a fertile and ecologically diverse agricultural landscape. Quite the reverse—*they are acting as nature acts.*

Through its reaction to such intervention, nature demonstrates just how apt this description is. In place of the relatively few forest species a great abundance of meadowland species, both plant and animal, arose, many of them species that were previously unknown in these latitudes because conditions were always unfavorable. Meadow plants which had been confined to marginal areas greatly extended their range, while a large number of immigrant species took up residence for the first time. The developmental impetus was so strong that numerous new plant species arose, either through cross-fertilization or through the colonization of fresh habitats by new mutants. In addition, the crop plants arrived, and with them a host of unintentionally introduced species. The same applies to the animals, particularly birds and insects (see also Suchantke 1993).

Since these early phases of human culture, however, conditions have changed radically. Whereas before nature itself provided the ideal landscape that sowed the first seeds of human development and set humankind on its path toward successful and far-reaching cooperation with nature, now in the temperate, largely forest-covered north nature appears to be dependent upon human action. At least it seems as if it is no longer capable of juvenilization, in other words, of being open to new evolutionary impulses, except through human intervention.

This is clearly apparent in the characteristic way ecosystems develop in this region. The same basic features appear again and again. C. Leuthold provides a particularly lucid example of this in his study

(1995) of the recolonization of the melt zone along the edge of the Aletsch glacier. An "embryonic" phase of occupation by plant microorganisms is quickly followed by the establishment of a varied herbaceous vegetation. There is no lack of woody forms, but they are small, either widely dispersed shrubs (willows) or young trees. There is plenty of space and light for the herbaceous plants, and this phase is characterized by a diverse insect life. It is all rather short-lived, however, as the growing trees soon displace the herbaceous vegetation, and closed forest establishes itself with an associated flora of woody shrubs—a plant community of few species, from which herbaceous growth forms have all but disappeared. In any temperate forest, regeneration will be found to involve a very similar succession—one need only think of the lush vegetation of a forest clearing, or the marsh phase in the sedimentation of a lake before it dries out and is taken over by forest.

When humans clear the forest, developing in its place a richly structured agricultural landscape, they are helping the juvenilizing tendency inherent in the plant kingdom to come to full realization at the ecosystem level. In thus checking the processes of ageing embodied in the forest and enabling nature to operate at a level that combines diversity with the greatest degree of openness to new developments, they are furthering the process of evolution itself. To what degree this can be described as "natural," that is, in keeping with nature, is readily apparent from the previously mentioned effect such human action has in encouraging the spread of local plants and animals, integrating immigrant species, and bringing about the emergence of new ones, both wild and domestic.

In a landscape where the juvenilizing attributes of the human and plant kingdoms have combined in this way, the contradiction between nature and culture is transcended. This shows also in the fact that in such a landscape there are no longer any fixed boundaries between wild and domesticated forms of plant and animal. Are the meadow grasses wild or cultivated plants? They are both! The woods are planted for commercial purposes, or are kept close to their natural state and carefully managed so as to contain trees of all age groups. Wild game is carefully preserved and controlled. But for the "artificial" dwellings of human settlements, redstarts, house martins, swifts, and swallows would have nowhere to build their nests. There is no end to this list. In the ideal cultivated landscape, all nature is cultivated and all culture is natural. And human beings are not only fully integrated into the scene, but play the key role in determining and preserving it.

Given all this, the fact nevertheless remains that the forest must not disappear completely. It is needed not only to maintain the wood supply, but also to preserve the natural integrity of the cultivated landscape: in every ecological community and in every population all age groups are always represented. This applies also to the whole evolutionary spectrum of living things: even to-day, and in spite of their temporal differences, we find archaic, early, later, and thoroughly "modern" forms all existing side by side. The old is endowed with the power to conserve, and as such is every bit as necessary as the new. The conserving character of the forests expresses itself in their ability to regulate humidity (and by implication the water cycle of the whole planet), thus exerting a balancing, stabilizing effect upon climate. In their absence we have been (and are) subject to extreme fluctuations of heat and cold, to drought and flood.

What has been presented here as a cultivated landscape is an ideal picture, which is now further off than ever before. That it could become reality, if we only had the will, is proved by the historical record. In culture after culture this ideal has been forged into one form or another, and some of these reached a high level of perfection—so high, indeed, that we are still living off them. There is some cause for hope in the fact that serious attempts to realize this ideal on a new level are becoming more numerous.

Old Land, Young Land

WE TEND TO THINK of New Zealand as a country full of sheep at the
edge of the world in the shadow of Australia. This is an attitude
ingrained in us during our school days by Eurocentric maps. Quite a
different picture is conveyed by the maps used by the children in, say,
a Japanese school. So why not for once follow the example of such
maps, set New Zealand in the center, and see what unaccustomed relationships come to light!

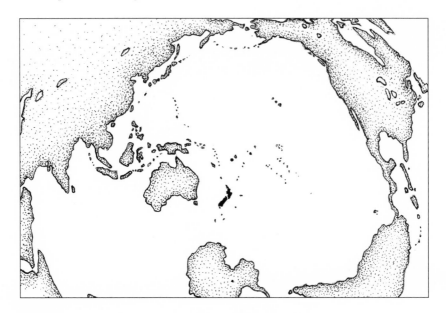

Fig. I. New Zealand seen for once at the center of the world.

GLOBAL CONNECTIONS

From this perspective the two islands appear in the center of a wide arc formed by the continents of the Old and New Worlds, with Australia still close by and Antarctica none too far away. In the immediate

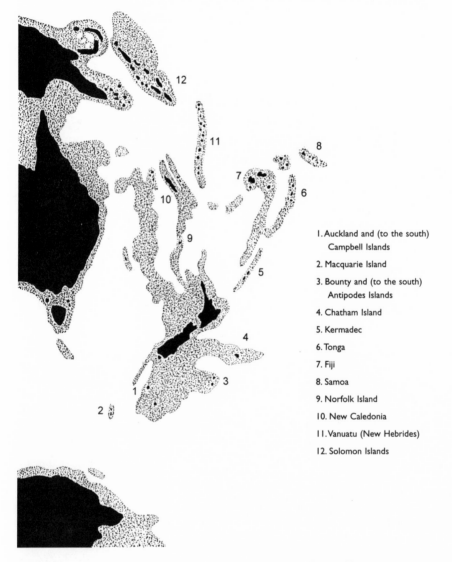

1. Auckland and (to the south) Campbell Islands

2. Macquarie Island

3. Bounty and (to the south) Antipodes Islands

4. Chatham Island

5. Kermadec

6. Tonga

7. Fiji

8. Samoa

9. Norfolk Island

10. New Caledonia

11. Vanuatu (New Hebrides)

12. Solomon Islands

Fig. 2. The submarine ridges and continental shelves around New Zealand. The shaded areas are above the 2000-meter level.

neighborhood further connections also become apparent, which are usually masked by "normal" maps. The shape of New Zealand itself bears witness to them. On the north island a long narrow spur points northwest, indicating the path of a series of underwater ridges, which

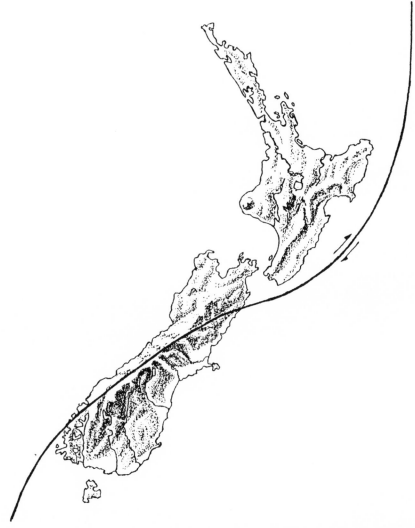

Fig. 3. The Pacific (to the east) and the Indo-Australian plates as their zone of contact appears in the context of New Zealand. On South Island the crest of the New Zealand Alps marks the zone of sub-duction and upward folding. Volcanic activity is strong on North Island, and evidence of it is found in the large, centrally located caldera of Lake Taupo, the two large, active volcanoes to the south of it, Tongariro and Ruapehu, and in Taranaki (Mount Egmont) on the island's southwestern headland.

in the past have undergone partial uplift.[1] Bypassing Australia, they run toward New Caledonia, New Guinea, and then on through the Sunda Islands in a great arc, which subsequently forges the link to the Asian continent. From the northeastern point of New Zealand a further line goes north toward Fiji and Samoa, surfacing briefly as the Kermadec Islands. These underwater ridges trail off to the south, forming only the Macquarie Islands on their interrupted way toward Antarctica.

Thus it becomes clear that there are connections with tropical Asia (and beyond that with the temperate lands of the Old World), with the Pacific islands, and with Antarctica. It is not immediately apparent, but there is also a connection with South America (Darlington 1969). In what follows the significance of these connections for a full understanding of New Zealand will be shown.

Australia, of course, figures large in the picture, and numerous related features are to be expected simply because of the shared past, as part of Gondwana, of these two neighbors. Nowadays, however, the character of Australia is very different. It is a continent of parching sunlight, of deserts, of acacia and eucalyptus savannas (unknown in New Zealand), where the drought never seems to end. Large tracts of it are covered with pyrophilous vegetation, so called because annual fires are essential to its life cycle (Walter 1968, 1983; Recher and Christensen 1981). Home to the marsupials and a diverse and colorful bird life, it is a continent of archaic cultures that never reached New Zealand. The contrasts with its smaller neighbor, however, were not always so fundamental, particularly in early Tertiary times, when a moist, subtropical climate prevailed, similar to that found today on the extreme east of the continent (Dawson 1988; Kemp 1981).

The "great arc" running up to the Sunda Islands is the result of the impact of two continental plates, the vast Pacific plate and the only slightly less vast Indo-Australian plate. In the region of large islands close to the Asian mainland the comparatively small Philippine plate is also involved. The situation of New Zealand is such that the contact line runs almost diagonally through the whole of South Island and

---

1. There is heated discussion in the literature about the actual timing of the appearance of such ridges. See Lovis 1989.

then passes east of North Island, thus placing it entirely on the Indo-Australian plate. The main peaks of the New Zealand Alps mark the contact line, while the whole range is the result of folding caused by the enormous opposing pressures involved. The situation is complicated by the fact that the plates meet at an angle, so that the westerly plate is raised and slides over the sinking easterly one (a so-called transform fault). This places New Zealand firmly in the volcanically and seismically active zone that spans the whole Pacific. The towering cones of Taranaki (Mt. Egmont) and the central volcanic plateau are prominent features of North Island, to say nothing of the many other centers of volcanic activity spread over the whole country. The recently increased activity of giant Ruapehu shows just how volatile the region is. Seventeen hundred years ago—comparatively recently in geological terms—vast tracts of North Island were smothered under massive layers of red-hot volcanic ash and pumice. It is not unusual for bulldozers involved, say, in road building, to unearth beds of charcoal that used to be forest.

That the contact between the plates was not always so straightforward as it is now is shown by the high degree of distortion to which New Zealand was subject after its separation from Gondwana-Australia. Only parts of South Island turn out to be Gondwanian rock, while the rest is a conglomerate of fragments transported there by plate movements and raised from the seabed—so-called terranes (Bacchus 1994; Cooper 1989; Lovis 1989).

STREAMS OF LIFE

Through the study of plate tectonics our understanding of the Earth has changed radically. The static picture of eternally fixed continents has been replaced by a dynamic one with a temporal dimension, a global process in which everything is in constant flowing motion (the rock cycle, as it were), with new structures arising while others disintegrate. Of course, in the former static view temporal processes were also recognized—volcanism, sedimentation, folding, etcetera—but these were all seen as local events, never as part of a global pattern bearing witness to the Earth as a differentiated, unified organism-like entity with its own inherent time structure. This convergence of geology with the life sciences has been bearing fruit for some time. The relationships between certain biological and geological processes have come to be more clearly perceived, and this has contributed to the elucidation of phenomena previously shrouded in obscurity. If one looks, for instance, at the distribution of living things over the Earth, considering it from the point of view of whole species or populations, then the long-term migratory patterns appear as flowing movements closely resembling the thermal currents of the Earth's mantel, which are the basis of continental drift.

In addition, the zones where plates are either touching or pulling apart are especially significant. Their features and function are gone into at some length in chapter 7 (see also Lattin 1967 and Suchantke 1982), so here it will only be necessary to sketch in what is relevant to an understanding of New Zealand. And we will confine ourselves, for reasons that will shortly become apparent, to convection zones— the lines along which two plates *meet*.

Under normal circumstances in such contact zones folding takes place. All the Earth's recently formed (alpine) mountain ranges arose in this way. This goes both for those that run north-south along the western edge of both Americas and for the predominantly east-west ranges of Eurasia, from the Alps to the Himalayas. As is well known, the latter form a well-nigh impassable barrier between different climatic zones, thus impeding the spread of many plant and animal species. *Along their length, however, they seem to play the opposite role by acting as migratory pathways.* Thus, after the last ice age central Europe was recolonized just as much by plants and animals from eastern Asia as by those emerging from the sanctuary of southern Europe. From the Korean-Manchurian region they spread in parallel to the great Eurasian mountain ranges, and this westward march has continued to this day. Exactly the same picture presents itself in the great western mountain ranges of the Americas, except here the axis runs north-south. The whole process becomes particularly dramatic where human populations join the scene, as was the case in Eurasia as well as in

North and South America. In the former case, the people of the eastern steppes, in following the mountain ranges, set the whole Eurasian
migrations in motion, while in the Americas the Indians expanded
their territories to north and south.

HOW DID ELEMENTS OF THE NORTHERN HEMISPHERE
END UP IN THE SOUTHERN ONE?

In New Zealand the observant visitor is constantly confronted by landscape features that are bewilderingly incompatible. Totally unspoiled
natural woodlands border areas of intensive farming where no indigenous plants are to be found, but in their place a conspicuous abundance of weeds, meadow and roadside plants from Europe and North
America: oxeye daisies, chicory, red clover, and countless others. All
around are heard the songs of birds that were once introduced, but
have long since become native: skylarks, blackbirds, chaffinches.
Opening one's eyes suddenly upon this landscape after having been
set down blindfold in the middle of it, one would have no notion of
its being tucked far away from Europe deep in the Southern Hemisphere.

But even in the wild, nature is likely to present one with sights
familiar from the temperate latitudes of the Northern Hemisphere; for
instance, four species of the pretty, golden-orange butterflies known
as coppers (*Lycaena*) are common (d'Abrera 1990; Gibbs 1980). How
can these members of a purely Holarctic (northern Eurasian–North
American) group, which does not occur anywhere in the tropical belt,
have found their way into the almost "butterfly-free zone" of New
Zealand? The feeling of treading familiar ground becomes particularly
strong on the upper slopes of New Zealand's mountains. Old friends
turn up at almost every step—eyebright (*Euphrasia*), forget-me not
(Myosotis), willow herb (*Epilobium*), gentian (*Gentiana*)—and these in
a considerable variety of species (Mark and Adams 1979). The single
most striking difference (of which more later) between these and their
European relatives is the universal prevalence of white flowers—there

Fig. 4. Two common New Zealand butterflies. Above: The golden shining male of the New Zealand copper *Lycaena salustius* and its darker female. Below: Tussock ringlet *Argyrophenga antipodum*. While resting in the grass it becomes almost invisible, the silvery stripes of the underside perfectly matching the structures of the environment.

are no deep blues, no radiant reds. In this these plants prove them-selves, very much in contrast to the ones that came in on the heels of agriculture, to be genuine New Zealanders. White is the dominant color of the indigenous flora. The question still remains of how these plants—all of them foreign to the tropics—managed to penetrate so far south.

The major clue lies in the lines made by the edges of the conti-nental plates, or, more exactly, by the *convection zones* where plates have collided and rock is being pushed both up and under. The line of contact between the Indo-Australian and Pacific plates meets the edges of both the Eurasian and the Philippine plates in the great region of peninsulas and islands along the southeast coast of Asia (see Fig. 5). The resulting convection lines, as investigation of the Malaysian flora has shown, provided the pathways along which the spread of the

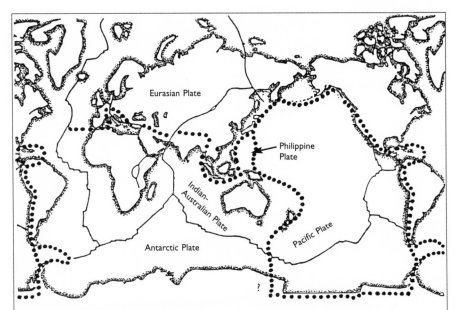

Fig. 5. Dilation zones worldwide (continuous lines), and compression zones (dotted lines), regions where collisions with consequent subduction and crustal folding are occurring. The extension of the compression line running through New Zealand to the south and the associated splitting of the Antarctic plate is hypothetical (after Schmutz 1986).

northern element took place! In the high uplands of Sumatra, Java, Borneo, the Philippines, and New Guinea, sparsely distributed remnant populations of various species of gentian, eyebright, and primrose (*Primula*) are to be found. They clearly mark two distinct routes: one is the Himalayan line running toward New Guinea through Burma, Malaysia, Sumatra, and the Sunda Islands; the other is the Chinese, which starts in China and heads southeast via the Philippines, northern Borneo and Sulawesi (George 1987; van Steenis 1964). These routes were taken not only by plants—the recent discovery in the alpine uplands of New Guinea of two close relatives of the copper butterfly has established the existence of a stopping-off point on the way to the far south (Sibatani 1974).

Of course, conditions in the tropics of today would not permit such a migration—the rainforests of the tropical lowlands present an impenetrable barrier to any organism from temperate regions. During the ice ages, however, the climate was considerably cooler, not only in the north but worldwide, and the rainforests were severely reduced in

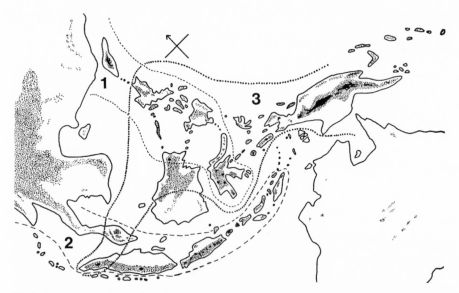

Fig. 6. Immigration pathways in the Southeast Asian Pacific according to the findings of plant geogra-phers: 1. the Chinese-Philippino line, 2. the Himalayas-Sumatra line; 3. pathway for the spread of species from the western and southern Pacific. (After van Steenis 1964.)

size. Conditions found today only in isolated pockets of the tropical uplands of Malaysia were then found considerably lower down and thus prevailed in much larger and more closely connected areas. It may be that they were much like the conditions found today in New Zealand: alpine vegetation, which in equatorial regions is confined to altitudes of between 3,000 and 4,000 meters, there extends down to 1,500 meters. Here the immigrants from cooler latitudes found con-genial conditions and extensive habitats, so that a great radiation of species took place. In the regions between north and south, however, they died out, except for a few stray pockets at extreme altitudes.

It should not be forgotten, or course, that the movement was not all one-way: certain plant species did ray out from the South Pacific and New Zealand to reach as far as the Sunda region and the Philip-pines. Some examples are a few primeval conifers of the araucaria or monkey puzzle family (kauri, *Agathis*), or the celery-topped pine (*Phyl-locladus*); also some members of the Liliaceae genus *Astelia,* which occurs in a wide variety of species in New Zealand, some of them epi-phytic, some ground-dwelling (Smith 1986).

THE SOUTH AMERICAN CONNECTION

In many places in the alpine landscape of New Zealand these botanical reminders of the temperate Eurasian north are joined by other impressions, which appear less in the presence of actual species and genera than in certain formative motifs that seem to affect both the shapes of plants and of landscape features.

To travelers in the high Alps of New Zealand, the dry and sparsely covered landscape on the off-wind, eastern side of the main watershed conveys the inescapable impression that they are in quite another part of the world—in the Andes. The gently rolling high plains with their bushy thickets and dry, yellow-brown grasses are strongly reminiscent of the summer landscape of the Peruvian puna, which exists at altitudes of 4,000 to 5,000 meters. New Zealand's "puna," much further south, is found at between 1,700 and 1,900 meters.

Even the details correspond. For instance, there is the cupressoid growth form of many dwarf shrubs (Poole and Adams 1990) with their closely packed, bud-scale-like leaves arranged in a four-edged formation (Fig. 7). Then there are the plants that appear as round, radiant white, rock-hard "cushions," some of them as much as a meter high,

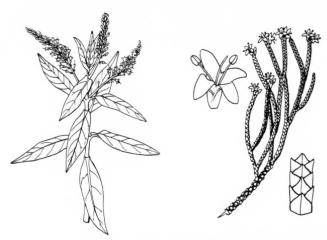

Fig. 7. Two members of the genus *Hebe*, close relatives of speedwell (*Veronica*). Left: the bushy *Hebe salicifolia*, which can grow up to 5 meters high and has long, willow-like leaves. Right: the dwarf *Hebe lycopodioides* from the New Zealand Alps with its scalelike (cupressoid) leaves. (From Poole and Adams 1990.)

huddled among the rocks, and known (for some strange reason) as "vegetable sheep" (Fig. 8). These are members of the daisy family (*Haastia, Raoulia*), whereas in the Andes the "vegetable sheep" effect is created by certain silver-white cushions of cacti (*Tephrocactus*) (Suchantke 1982). There are also other "cushions"; these are flat and look like spread-out cloths. Significantly enough, many of these will be found to belong to a species that has close relatives in South America: *Phyllachne colensoi* of the small family of the Stylidiaceae. These emerald-green cloths are strewn with myriad tiny blossoms. Also the prize of New Zealand's alpine flora, the large-blossomed yellow and white crowfoot species *Ranunculus lyallii* (Fig. 9), *buchananii, nivicola, haastii,* and others, display more of a morphological resemblance to Andean species than to their indigenous lowland relatives (Fisher 1965).

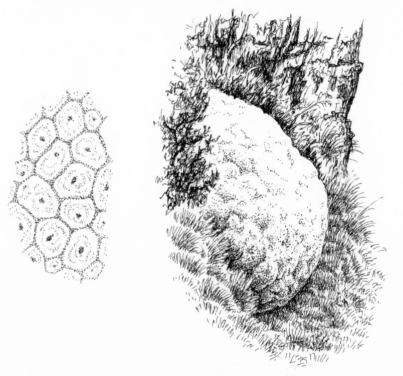

Fig. 8. A "vegetable sheep," the hard white cushion, almost 2 meters in diameter, of *Raoulia eximia*, from the subalpine region. Left: a detail—all that can be seen of the branches are the tips, which appear as dark points; their leaves are hidden inside the white (transparent!) wool. St. Arnaud's Range, Nelson, South Island, January 1995.

Fig. 9. "Mount Cook lily," in reality a white-flowering buttercup (*Ranunculus lyallii*), easily reaching 1.5 meters in height. The popular name seems quite appropriate for the splendor of this big-flowered, high-growing plant of the New Zealand Alps.

New Zealand's connection with Patagonia ("Fuegia") and the Falkland Islands is particularly close: "It is a fact that many plants and invertebrate animals of the southern tip of South America (with Tierra del Fuego), the southern corner of Australia (with Tasmania) and New Zealand are related" (Darlington 1969). There are, for instance, significant overlaps in the Geometridae, a family of moths, and, among the plants, chiefly in the southern beeches (*Nothofagus*). But there are also the three New Zealand Fuchsia species, among them a very common and familiar one with trunk and branches twined and twisted, with reddish, peeling bark and small inconspicuous blossoms: *Fuchsia excorticata*. In South America the wide range of Fuchsia species are a characteristic feature of the cloud forests on the eastern slopes of the Andes, which also feature large numbers of giant ferns and are climatically very reminiscent of New Zealand. There are even some exceptional instances, in which the same plant species occurs both in New Zealand and in Patagonia—*Hebe salicifolia* in Fig. 7 is an example of this.

If one looks at the distance between the tips of the southern continents on a globe rather than a map (on which the polar regions tend
to be extremely stretched), they do not appear so terribly far apart.
Seeds can be distributed by ocean currents or birds. More to the
point, however, is the fact that for parts of the Tertiary period the edges
of Antarctica had plant colonies. This, together with the fact that most
of the plants common to South America and New Zealand are associated with cool temperate climates, would argue for migration via
Antarctica.

A hypothetical continent of "Pacifica" seems, according to oceanog-
raphers, never to have existed, as the floor of the Pacific shows no
trace of sunken continental crust (Lovis 1989). A further hypothesis
(Schmutz 1986) envisages the convection zone that runs through New
Zealand continuing on along the Antarctic plate, turning eastward over
the Antarctic Peninsula, and running all the way to Patagonia.
Although this might be regarded as a purely speculative assumption
made because it fits in with the Earth's tetrahedral structure as mapped
by Schmutz, it surely finds corroboration in the biogeographical connections between New Zealand and South America. They clearly imply
the existence of a line of convection acting, in the manner of all such
lines on land, as a thoroughfare of biological migration between these
two regions.

BEECH GROVES AND TROPICAL RAINFORESTS

The landscapes of New Zealand continually present the visitor with
new riddles. How is it possible that vegetation types normally continents apart are here found cheek by jowl in mosaic patterns? If one
has just been in a sunlit beech wood that could easily have been somewhere in central Europe, it comes as quite a surprise to find oneself
not much later in the gloom of a thickly tangled jungle that feels distinctly equatorial.

It is the overall impression of the southern beech woods that makes
them seem European, and especially the quality of the light streaming

Fig. 10. Symbols of the "cold tropics": snow-capped tree ferns in southern winter, mid-August, near Lake Waikaremoana, North Island.

through the great palatial spaces suffused by the emerald glow of the sunlit foliage. On more detailed observation the picture changes. Many of the tree trunks fan out at the base in a way reminiscent of buttressed tropical forest trees. The leaves are small and look identical to those of the northern scrub birch (*Betula glandulosa*). They are leathery, tough, and, unlike birch leaves, evergreen. Impossible to ignore also are the numerous epiphytes, especially the tree ferns, which lend the forest a certain aesthetic magic. In winter, however, when they are capped with snow, they look sad and out-of-place.

In all probability the southern beeches are not immigrants from the north, although it would seem reasonable to think they were. After all, the family of the *Fagaceae*—to which the oaks and chestnuts also belong—enjoys it greatest abundance and species diversity in the Northern Hemisphere, some genera even penetrating over the equator as far south as New Guinea, the northern limits of southern beech distribution. Nevertheless, no fossil pollen of *Nothofagus* has ever been found in the Northern Hemisphere, and if fossil pollen finds are to be trusted it would appear that the southern beech emerges earlier in the Earth's history than its northern relatives of the genus *Fagus*. Its occurrence in a variety of species all around the lower Southern Hemisphere, together with pollen finds on the Antarctic continent itself, would argue for a southern origin. Quite possibly the ancestral group from which the beeches sprang split into a northern and a southern branch, each of which underwent a separate development as Gondwana and Laurasia (the great northern continental mass) drifted apart (George 1987).

THE "COLD TROPICS" OF NEW ZEALAND

A stronger contrast to the beech woods than that presented by the other type of New Zealand forest could scarcely be imagined. In place of open, light-filled spaces we now have the jumble of lianas and long trailing rhizomes of the ferns that climb the tree trunks, at least insofar as the mosses and lichens allow them a foothold. Delicate filmy ferns cover the trunks of the tree ferns in a thick layer. High above in the not-too-dense canopy sit massive clumps of epiphytic lilies (*Astelia*), while from the branches hang long filaments of club mosses and white bunches of tender orchid blossoms. Trees follow the example of such growths to reach the light. Deposited by birds, their seeds sprout in the nestlike clumps of the astelias, and send long root strands down the trunk of the host tree. While the crown of the young tree unfolds above, the twining, liana-like roots thicken; in some species they merge, thus strangling the host tree and eventually taking its place. This form of growth is known in all the Earth's tropical forests, and while it is usually seen in various kinds of fig tree (strangler figs), in New Zealand it appears in other plant families entirely: for example, various members of the genus *Metrosideros,* which belongs to the Myrtaceae and is therefore related to the eucalyptus (which does not occur in New Zealand), and *Griselina,* which, like dogwoods and bunchberries, is among the Cornaceae (Dawson 1988).

Fig. 11. An old Matai tree *Podocarpus spicatus*, with its trunk completely covered by the liana-like roots of various bushy and treelike "squatter" species, plus numerous climbers and epiphytes. Lake Kaniere, Westland, South Island, December 1994.

This vegetation type attains its climax as swamp forest on the west coast of South Island, the wettest area of New Zealand. The hordes of epiphytes, the thick carpets of mosses and ferns on the tree trunks, and the endless tangle of lianas and creepers create a picture that is comparable only to the rampant hypertrophic growth found in hot and humid coastal regions of the tropics. This picture is at first baffling—we are, after all, in a cool, temperate region. However, the abundant rainfall and constant humidity carried in off the sea by the prevailing west wind prevent any chance of growth-inhibiting frost.

Fig. 12. Swamp forest in the "cool tropics" of South Island. Immature and mature forms of the Kahikatea tree *Dacrycarpus (Podocarpus) dacrydioides*. In the foreground large rosettes of *Phormium tenax* ("Flax"). Drawing by Nancy Adams (Poole and Adams 1990).

In these forests of the "cold tropics" the trees are no longer broad-leaved, but rather conifers of the Southern Hemisphere. They belong to the yew family (Podocarpaceae) and are phylogenetically much older than the southern beeches, which figure among the angiospermous flowering plants. There can be no doubt that they are an ancient Gondwanian feature, since they exist today on all the landmasses that were once part of Gondwana: Australia, India, Africa, and South America. In these last two their range extends north far over the equator, while in East Asia it even exceeds its Gondwanian limits.

One glance at these trees is enough to reveal that they have little in common with conifers of the Northern Hemisphere, showing no trace of the geometric regularity of a spruce or a young fir. Furthermore, their great diversity of shape and their tendency to look very different as saplings from the way they do in maturity only serve to increase the confusion of any visitor wishing to learn how to distinguish them.

From a distance a podocarpus forest appears olive or brownish-gray, totally different from the light green of the southern beech woods. Should rain clouds pass overhead, everything sinks into deepening tones of gray. There are, of course, virtually no colored blossoms. One feels transported back to very early times in the Earth's prehistory, when the plant world was dominated by nonflowering species and, although some of the first conifers bore flowers (gymnosperms), full-blown, colorful flower heads were unknown.

A little color appears, if somewhat shyly, in clearings and at the forest margins in the form of emerald-green tree ferns and the muted red of the tall blossom stalks of New Zealand flax (*Phormium tenax*), an agave-like member of the lily family, so called because of its tough fibers that can be used to make ropes.

The very pinnacle of tropical vegetation in New Zealand is found on the northern half of North Island. Here there are almost pure stands of the mightiest of New Zealand's trees, the kauri (*Agathis*). It is a relative of araucaria (monkey puzzle), but looks quite different: gray trunks of truly gigantic proportions rear up into the sky like great ramparts. They have widely spreading branches that form a bushy but not too dense canopy, in which whole gardens of epiphytes can be accommodated without even showing. Unfortunately, in the past these venerable giants have been subjected to such rapacious logging that not many of them are left. In Waipoua Forest stands the mightiest of them, *Te Matua Ngahere,* the two-thousand-year-old "father of the forest." To this day this tree is revered by the Maoris and stands under their special protection. In New Zealand at present, as in many parts of the world, there is a sense of alarm at the destruction that has gone on, and this has led to vigorous attempts to save and care for what is left. The way these three vegetation types—the southern beech, podocarpus, and kauri forests—exist side-by-side in a mosaic-like pattern is one of the essential and most striking features of New Zealand. Its unique significance resides in the fact that plant communities from different periods in the Earth's history thus inhabit the same moment. The "cold rainforests" undoubtedly belong to older (Cretaceous) epochs when the climate was considerably warmer. The life that existed at that time on Antarctica adds weight to this view. The spread of the southern beeches, by contrast, is of more recent date, as their still persisting tendency to encroach upon the territory of the podocarpus forests would seem to indicate. They are younger, more "up-to-date" with their closer adaptation to current climatic conditions (Walter 1968).

Of course, we must also include in the picture the vegetation of the high uplands with its contingent of northern migrants. It is younger still, bearing as it does the marks of the last ice age. And if we then add in the farmland plants brought in almost yesterday on the heels of agriculture, we have an incredible range of developmental phases that followed each other in time, but here share the same geographical stage. It might be objected that, far from being unusual, this is a commonplace, since everywhere in nature ancient and more recent forms of life are found in close proximity. A case in point might be an organism such as the earthworm, which is known to have existed from very ancient times, being eaten by another organism that has only been around since Tertiary times, the robin. But that is surely something other than the simultaneous coexistence of plant communities, each of a different age and with a wide distribution. What makes New Zealand special is the fact that on its soil several complete landscape types, ecological "provinces," meet, which together document the development of this "microcontinent."

THE INFLUENCE OF THE SOUTHERN OCEANS

While it is true that on a world scale tropical elements pushed southward in an unparalleled fashion, it is also the case that they were met by antarctic and subantarctic influences emanating from the surrounding oceans. Whereas New Zealand's North Island puts out great tongues of land into the ocean, the situation is reversed in the south, where, through narrow fjords, the ocean invades right into South Island's interior. These are the haunts of dolphins and of penguins, which breed under the branches of the coastal beech forests. In the interplay of sea and land the polar element in the natural history of New Zealand reaches its high point: subantarctic ocean meets on land an ancient tropical mode of life. The result is not a feud, but enhancement. Without the influence of the sea the tropical luxuriance we so much admire would not be possible. This is most clearly apparent on South Island. To the west of the Alpine watershed the high rainfall coming in on the west winds creates the lush vegetation of these "cold tropical" regions, while east of the mountains lie dry grassy plains with a much more continental climate. The ocean's effect of moderating temperatures enables the survival of the formerly warm, currently cold tropics as a living anachronism.

The tremendous abundance of life in the southern oceans is reflected in its vast numbers of fish, which in turn are the cause for the fact that there are many species and large populations of dolphins, seals, sea lions, and, above all, sea birds. The seal colonies on the coastal cliffs were once much larger, but are gradually beginning to recover now that they are protected. Several species of penguins breed on these coasts, and there is a breeding colony of royal albatrosses near Dunedin. Attempting to name every species of petrel, cormorant, gull, tern, plover, and oystercatcher that this coastline and its waters support, not to mention the many breeding and nonbreeding guests from the subantarctic islands (and, during the northern winter, from the East Siberian Arctic), would make for a long list. A good glimpse into the underwater world is provided by a walk along the shore at low tide. In rocky places the brown seaweed *Macrocystis* hangs limp, waiting for the tide to turn it once more into an underwater forest, while washed up on sandy beaches is an inexhaustible variety of red seaweeds in various shades of scarlet and crimson, with forms ranging from the delicate and feathery to the broad and ragged. At a spring tide it can happen that mudflats in bays left suddenly high and dry are dyed red by millions of *Munida* shrimps all jumping about in desperation. Vast shoals of these, just like the famous krill (*Ephausia*) in antarctic waters, play a vital role in subantarctic seas as the main food of whales.

BIRDS IN PLACE OF MAMMALS

This great abundance of animal life persisting today in the surrounding oceans and along the coasts stands in striking contrast to the paucity of species on land. There is no scarcity of the lower animals, those vigorous "plant-like" creatures involved in processes of vegetative growth, regeneration, and reproduction. It is among the higher animals, those endowed with inward sentience, that the pinch is felt— the birds and especially the mammals. This peculiarity of New Zealand becomes really clear if the state of things there is compared with that of its not-too-distant neighbor Australia. Australia is also a land that the more highly developed, "modern" mammals failed to reach, but there a richly diverse marsupial fauna developed in their place, as it were. Nothing of the sort occurred in New Zealand. The only mammals existing there before the arrival of humankind were two species of bats.

In contrast to mammals, there appears to be no lack of birds. But for color and sheer variety New Zealand cannot compare with Australia, with its cockatoos and parakeets, kingfishers and bee-eaters, its lyrebirds and numerous singing birds. At first glance New Zealand's bird life is extremely unspectacular. The voices of its indigenous species are very unprepossessing and tend to be drowned out by the songs of immigrant blackbirds, song thrushes, chaffinches and larks, which can be heard literally everywhere, not only in landscapes modified by humans. That the newcomers were able to spread like this may well be due to the presence of a large number of unfilled ecological niches, but undoubtedly also to the proven decimation, and in some cases complete elimination, of indigenous species by imported rats and feral domestic cats. The Europeans are adapted to life with predatory mammals, but not the New Zealanders!

Nothing could indeed be stranger than the paths of development taken by New Zealand's own bird life. Since mammals were absent until the arrival of humans, some of the ecological niches that they would normally fill were taken up by birds; or, to put it another way, certain integral parts of the total ecological organism—certain of its organs, if you will—which elsewhere would be represented by mammals, in New Zealand take the form of flightless birds. This is one of nature's most widespread phenomena: if an ecosystem comes to "require" an organism with certain functional features, or, to put it another way, if a certain ecological response becomes necessary within a given set of conditions, it will always be found to have been developed in whatever is locally at hand. For instance, arid regions of the

Fig. 13. A giant moa *Disornis giganteus* and some kiwis. (From Hochstetter 1867.)

world call for some life-form that is "cactus-like," able to guard against water loss by reducing its leaves and thickening its stem into a storage organ. In America it is the "actual" Cactaceae that fulfill this role, whereas in the Old World, where there are no cacti, this is done by members of the Euphorbiaceae, the spurge family.

Thus in New Zealand the role of grazing animal, filled in other continents by hoofed animals and in Australia by large kangaroos, was played by flightless birds related to ostriches, the moas. There used to be numerous species of these in all sizes up to giant forms of five meters in (head) height—from "deer" to "elephant," as it were. They died out with the coming of humans; but their natural demise might already have begun before this time. Surviving today, although under serious threat in some quarters from imported mammals (mainly rats, feral domestic cats, polecats, and weasels), are the kiwis, the kakapo (a large nocturnal flightless parrot), and two equally flightless rails: the goose-sized takahe and the smaller weka. They all live in the cover of woods and thickets or in the tall grass of the mountains (takahe) where they can dart about with great agility, just like smallish mammals living in similar habitats.

Not to be forgotten in this company of strange creatures are two species of parrot, which, even though they fly very well, also "stand in" for a group of mammals: the monkeys. The kea and the kaka (the Maori names imitate their calls) are the clowns among the birds of New Zealand. Keas can spend hours in tussles and games of one kind or another, and they have long ago acquired the habit, just like baboons and guenons in Africa, of lying in wait for tourists at stopping places. In New Zealand's ski resorts they ambush visitors in true monkey style, and can be quite a nuisance. Intelligent and full of curiosity, they will remove all of a car's cables and rubber seals if the owner is not careful. Like many monkeys they also have the slightly macabre inclination to eat meat when they get the chance; just as baboons will prey on young antelopes, keas attack grazing sheep and tear out lumps of flesh. While the keas are seasoned inhabitants of the mountains, their cousins the kakas live in lowland forests. The way they rouse themselves in the morning with deafening screeches while flitting wildly through the treetops is remarkably reminiscent of the way troupes of capuchin monkeys or guenons romp around with loud cries, "flying" from tree to tree.

This is a truly fascinating phenomenon, which, surprising as it may seem, has so far been glossed over by science with the "explanation" of evolutionary convergence. How does it happen that when the evolutionary level of mammal has been reached in a region where there are no mammals, other animals of the locality—in this case birds—take on mammalian characteristics? This does not fit in with any conventional theories of evolution, which all share the assumption of blind, random mutation. This phenomenon clearly implies the operation of some higher developmental principle that has nothing to do with genetic kinship, but instead—to echo the phrase used previously—is able to make the most of whatever is locally at hand. The large ecological organism of which all these animals are integral parts might be regarded as the decisive factor here, but the morphological effects of which it is capable are not enough to account for this phenomenon. Most likely both are involved, the one working through the other—but we have no knowledge of how this happens nor of the nature of these formative influences, for the simple reason that so far no attention has been paid to them.

The fact that in addition to covering their own roles in the ecological drama the birds had to act as stand-ins for the mammals is all the more remarkable when we consider that the bird life of New Zealand is itself somewhat deficient. As already mentioned, many of the ecological niches elsewhere filled by birds are empty—a phenomenon characteristic of remote islands. Such unfilled niches are an open invitation either to local organisms that happen to be suitable, or to new arrivals. New Zealand seems to be a veritable magnet for immigrant species, which are drawn there in a continual stream that shows no sign of letting up. For instance, there were no swallows, apart from a few occasional visitants. Then in 1958 the first brood of the Australian welcome swallow was registered, and today it is found everywhere on both islands. Further recent examples are the white-faced heron (first recorded 1941), the black-fronted dotterel and the spur-winged plover (both 1950s), all Australian species, and now common throughout New Zealand (Chambers 1989; Falla et. al. 1980). The European settlers did their bit by bringing in the missing songsters and garden species. In a very short time these had spread to every corner of the land. Visitors from Europe find it somewhat disorientating at first to be walking in a tropical jungle with the familiar sounds of a northern spring in their ears—the flowing melodies of blackbirds, the trilling of chaffinches, and the fanfares of song thrushes.

Whereas the introduced and immigrant birds have settled in very smoothly and provided a wonderfully musical enhancement of the indigenous bird life without causing any displacements, this cannot be said of the introduced mammals. But before getting to this there is a little more to be said on the subject of deficiencies.

FLOWERS WITHOUT COLORS

For anyone stepping for the first time on New Zealand soil, the flowers are a sad disappointment—they are all white. There is the occasional bit of yellow or a touch of blue or red, but always only the merest touch. Colored blossoms appear almost exclusively on introduced garden plants and others that have escaped from the gardens.

It is on the first sally up the high mountains, with equally high expectations, that this hits home the hardest. From experience of other parts of the world one is used to encountering the strongest colors on the high slopes. In New Zealand the gentians and willow herbs, forget-me-nots, alpine avens, and many others well known for their blues, reds, and yellows in the high mountain ranges of the Northern Hemisphere are all just as white as exclusively local native species. Members of families that have colored blossoms in Australia and Tasmania are also white. It appears reasonable to connect this lack of floral color with the corresponding lack of pollinators such as butterflies and bees, for which colors act as signals. This assumption must be treated with caution, however, as is shown by the fact that on the subantarctic islands (Auckland, Campbell, Antipodes, Macquarie) there are plants that flower in a riot of purple in spite of the lack of insects (*Pleurophyllum,* a genus confined to these islands) (Dawson 1988).

It is nevertheless impossible to miss the fact that not only the colors but also the forms of blossoms are intimately connected to their pollinators. Visitors coming from Europe to North America for the first time, for instance, are likely to be struck by the intense, even aggressive, scarlet of some blossoms—for example, those of cardinal flower, scarlet larkspur, catchfly (fire pink), and many others—a color not found on any European plant. If they then travel on to Central and South America, they will encounter such fiery tints on all sides and on a wide variety of plants, whose diversity is matched by an increase in the variety of hummingbirds, their main pollinators. And it is not only in their colors that the flowers are "designed" to attract the attention of these visitors, but also in their forms. Bird-pollinated flowers usually have petals arranged in a long trumpet shape, out of which the stamens and stigmas protrude far enough to ensure that when a bird inserts its beak its brow will be dusted with pollen or the stigma will receive pollen from another flower.

A well-known example of such South American flower forms is the scarlet sage, used to create a riot of red in many a European garden. Bees trying to get at the nectar in the flowers of this plant always fail, for it is tucked away out of reach of the bee's proboscis, and anyway there is no landing place for insect visitors (Fig. 14). With European (and some North American) sage species the situation is quite different. Their flowers are not very striking visually, being delicate blue or pink in color, but the most important difference is in their form. A broad lip provides the bee with a comfortable landing place, and, on sticking its proboscis into the (shallow) flower, it first encounters a sort of trapdoor, which presents no obstacle as it opens very easily. It is joined to the two long stamens, which lie in the blossom's "hood," but are bent downward when the "trapdoor" is opened, thus brushing the bee's back with pollen. When the stamens wilt in the course of development, the style grows downward out of the hood (Fig. 14). The stigma at its tip will touch any visiting bee, and thus be pollinated.

From where do the flowers involved in such relationships take their form? In all likelihood the answer appears to be: from the animal—insect or bird. Bees are not dependent upon sage blossoms; they visit a whole range of plants. Without bees, however, sage would not produce seeds and thus could not survive. Its very existence presupposes that of bees. Its whole morphological history is thus intimately associated with the bee, which is the source of the formative influences! The case of plants pollinated by hummingbirds is just the same, nor are things any different with those species which in New Zealand have inconspicuous white blossoms and in Australia colored ones—for instance, members of the southern heath family (*Epacridaceae*). In Australia the dominant pollinator role is played by a bird family, the honeyeaters (*Melliphagidae*), of which New Zealand has only three to four species, all confined to the lowlands.

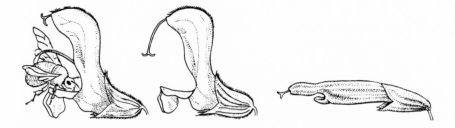

Fig. 14. European sage (left, middle); scarlet sage.

By contrast, the mountain plants of New Zealand typically have unspecialized moth-pollinated flowers, and their white blossoms show up well in twilight conditions; many of them also give off a strong scent. To be abroad in the evening, or even in the daytime when it is very dull and cloudy, is to observe on all sides the tiny "butterflies of the night" fluttering busily round the blossoms. They are found everywhere in large numbers, both of individuals and of species, and fill the role (together with flies) of the "butterflies of the day" and the bees.

A number of important insights arise out of all this. First, it becomes clear that the evolution of living things is always a product of the action of two antagonistic or complementary influences, one internal and one external. The former is the basic inherited form, the ground plan, the "type," which the given plant or animal has in common with its genetic relatives. The latter, environmental influences such as climatic, light, and soil conditions, bring about changes of form, color, etcetera, and in the case of plants are especially important in determining the growth form of stems and leaves. For instance, as previously mentioned, the "cactus" growth form is found in all the Earth's hot deserts, but in Africa it has been adopted by a totally different plant family from the one that follows the same pattern in South America. In the case of blossoms the decisive formative influences affecting shape and color emanate from the pollinating animals. As we saw, in many cases these are so strong that blossom and pollinator appear like two parts of one whole, while in all cases at the very least the form and function of the blossom only make sense in connection with a particular pollinator. The only word for this is *coevolution.*

In this way the flower receives something that initially has nothing to do with its basic nature, its "type," and that is directed through certain features of shape, color, and scent toward beings of a sensory nature, i.e., animals; thus the flower uses the same mode of expression that animals themselves use to express what is going on inside them. Animals have at their disposal a rich spectrum of expressive possibilities in terms of body language, color, and olfactory signals. Plants, on the other hand, possess neither a sentient "inner life" (and by implication the possibility of expressing such sentience), nor sense organs with which to perceive its activity. Thus everything about them that belongs to this sensory realm is purely an image that quite clearly originated in their counterpart—the insect or bird—and has been impressed upon them from the outside. In form and color the flower is an image of the instinctual inner nature (as well as the bodily structures) of its pollinating visitor. It is an *image of sentience,* but not of its own sentience, which is entirely lacking.

Thus in the colorlessness of New Zealand's flowers something we have already remarked upon is repeated in image form, namely the dearth of higher animals. This was apparent in the absence of mammals and the sparse bird life, also indeed in the relatively late appearance of humankind. Now it appears again in the dearth of colorful insect life (especially butterflies) and the lack of flower color among the flora.

This all belongs, more or less, to one side of the picture. The other side has also been repeatedly referred to. It is expressed in the diversity, luxuriance, and vitality in the vegetative aspect of New Zealand's plant life. The one is strongly dependent upon the other.

It is a well-known phenomenon in the evolution of animals that as neurosensory organization increases in complexity of structure, function, and "internal" activity, vegetative capacity diminishes. The faculty of vegetative reproduction is lost, together with the ability—widespread among the lower animals—to regenerate limbs or grow whole organisms from single fragments. We are all too familiar with this polarity through our own mental life, in which the degenerative effects of the normal waking state must be compensated for by regular periods of unconsciousness, when the vegetative side of our organism is able to regenerate itself. Plants are spared such problems. In the evolution of the plant kingdom we do not see the gradual development of an "internal" life. As already mentioned, what becomes "inward" in animals retains its peripheral nature in plants, so that there is no question of a reduction, let alone a loss, of vegetative potential. Even the most highly developed flowering plants fully retain the faculties of regeneration and vegetative reproduction.

PRIMEVAL VEGETATIVE ABUNDANCE

Among the special features of New Zealand is the fact that its poor showing in the way of blossoms, the highest expression of plant form, is counterbalanced by a tremendous abundance and morphological diversity in lower nonflowering plants. The *seaweeds* around the coasts have already been mentioned, and the astonishing variety of form they display is amply demonstrated in a recent, richly illustrated book (Adams 1994). And as for the *mosses*—one has to come to New Zealand to discover just how variable they can be. The moist climate creates ideal conditions for them. Along streams in the forest gloom, liverworts cover the stones with the dark-green oily sheen of their flat amorphous "leaflets," which at certain times are decked with a delicate filigree of little umbrella-like structures (Fig. 15). In upland areas of particularly high rainfall the branches of the trees that throng the slopes of mountains and volcanoes are festooned in thick moss. Decorative mosaics of bright green stars, reminiscent of blossoms, are strewn over the forest floor, interspersed with dark green miniature palm groves. Less common is the giant moss *Dawsonia superba*. It is of truly gigantic proportions, sometimes reaching a height of half a meter and looking looks like a young fir tree (Fig. 16). On dry upper slopes of volcanoes lie what look like patches of melting snow, spread out in

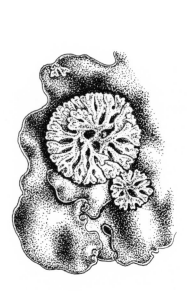

Fig. 15. Left: Over the deep-green, formless lobes of the Liverwort *Forstera* stand the delicate pale-green, flower-like structures of its organs of sexual reproduction. Right: Yellow-green fronds of a moss species (*Hypnodendron*).

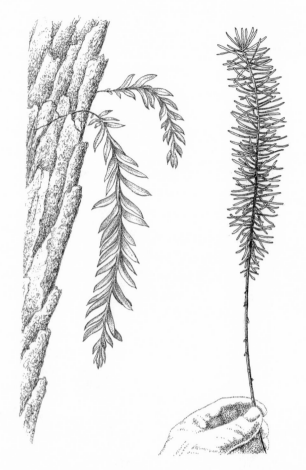

Fig. 16. The giant moss *Dawsonia superba* (right) and the unassuming, but nonetheless interesting *Tmesipteris elongata,* one of the oldest indigenous land plants, growing on the trunk of a tree fern between the scalelike stumps of fallen fronds.

stark contrast to the black of the volcanic boulders all around. These are the hard, extremely resilient white cushions of the moss *Rhacomitrum pruinosum*. Everywhere the forest trees are draped in *lichens* in a strange tangle of "antlers" and shapeless sheets. Often enough they compete with the carpets of moss, the one overgrowing the other. The lichen flora of New Zealand is "one of the richest and most interesting in the world" (Galloway 1985).

Another group having a seemingly inexhaustible variety of form is the *ferns* (Brownsey and Smith-Dodsworth 1989). Whereas in most northern countries they lead a fairly marginal existence, in New Zealand they determine the character of the landscape, mainly through the presence of their largest representatives, the tree ferns (see Fig. 10). Except for the very driest areas, these are encountered everywhere, in every clearing, in the forest undergrowth, along the banks of streams, and beside roads. To see them is always an aesthetic delight. Their geometric regularity has a composite beauty, whether this appears in the mesh of their scissor-like fronds against the sky, or in the filigree of bright green when the sun shines through. The more modestly proportioned ferns also abound, of course; there is scarcely a landscape that lacks them. The forest floor is in many places one solid carpet of fern fronds, while the same funnel-like shapes grace the branches and trunks of trees wherever the moss leaves them room. Among them are epiphytic and climbing species, as well as liana-like creepers. Every imaginable form from the temperate lands to the tropics is present (Fig. 17). The *club mosses* display a similar range of form, although in number of species and density of population they cannot match the ferns by a long shot, as indeed is the case all over the world. They hang as natural curtains high up in the trees, clamber about in the bushes with long multibranching shoots, or stand like elegant miniature fir trees on the forest floor.

Horsetails (*Equisetum*) are not part of this picture. This is evident from the field horsetail (*Equisetum arvense*), an importee found only on farmland and in neglected corners—an interesting and, in its own way, telling phenomenon. Within the fern family (*Pteridophyta*) horsetails represent the opposite pole to the actual ferns. Whereas the ferns go completely into the formation of large leaves, thereby branching out into an abundance of different forms seemingly inexhaustible in terms of size and sheer vegetative mass, the horsetails are reduced merely to the formation of shoots, all leaflike structures being suppressed. They are thus to a considerable extent an expression of influences that inhibit the vegetative, and these are not very strongly present in New Zealand. The club mosses take a similar path, but do not go quite so far, for they still have scalelike leaflets. It is significant that *Lycopodium volubile*, a particularly common species, arranges its

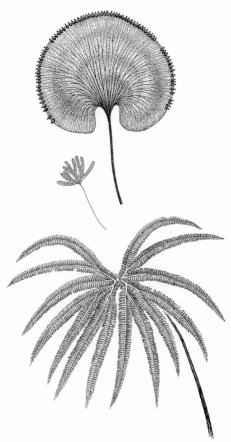

Fig. 17. Unusual forms of New Zealand ferns. Above: *Trichomanes radicans* ("kidney fern")—the fronds are edged with a thick line of spores (up to 10 centimeters in diameter). Beside it the inconspicuous *Trichomanes endlicherianum*, which forms mosslike carpets on the trunks of tree ferns. Below: the considerably larger *Sticherus cunninghamii* (umbrella fern), whose decorative fronds can reach a length of 1 meter.

branched side shoots into one flat plane in the manner of a leaf, thus approaching the growth form of a fern.

Of special interest is the occurrence of primeval distant relatives of the ferns from the genus *Tmesipteris*, members of the family *Psilotaceae* (Fig. 16), which can be found as epiphytes on the trunks of tree ferns. They add a further dimension to the picture of a plant world with strong primeval features, to which the ancient Podocarpaceae and the kauris belong. The animal world fits the same pattern. The unremarkable, but all the more famous *Peripatus* (Fig. 18), which lives hid-

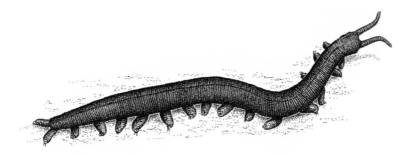

Fig. 18. Not a pretty sight, but an interesting one: *Peripatus*, 2–5 centimeters long, a link between ringed worm and arthropod.

den in caves and rotting wood and is an ancient Gondwanian element that occurs in all southern continents, has also been preserved in New Zealand in numerous species (Ruhberg 1985). Its fame is based upon the fact that it looks like an annelid worm with lots of rudimentary legs, and thus, simply speaking, represents a transitional form between earthworm and insect.

Primeval forms completely restricted to New Zealand are also found among the reptiles—the tuatara—and among the birds. The tiny New Zealand wren (Acanthisittidae), which due to its very short tail looks like a ball of feathers, is regarded as the most primitive of the sparrow family (Fig. 19) (Sibley et al. 1982).

Fig. 19. Yellowhammer? Warbler? Tree creeper? The tiny "Rifleman" *Acanthisitta chloris*, one of the archaic family of New Zealand wrens. Its English name comes from the resemblance of its green, yellow, and black markings to the uniforms of the British colonial army. After drawings by David Cemmick (Cemmick and Veitch 1987).

NEW ZEALAND: RESERVOIR OF FORMATIVE POTENTIAL
FOR THE FUTURE?

It would certainly be wrong to regard New Zealand as primarily a reserve or retreat for primeval organisms, both plant and animal. Nevertheless, when the underdevelopment of higher animal life, the deficits among the flowering plants, and the significant role played by lower forms, especially those of the plant kingdom, are all viewed together, there is a strong inclination to do so. *Of course, to be archaic, to remain at a primeval level of development, could also mean not to have used up the available evolutionary potential.*

New Zealand simply abounds with evidence of this. Besides the astonishing morphological variety among lower plants, there are among the higher plants in New Zealand many woody species, in which single individuals are found to display a wide range of different leaf-forms—a somewhat unusual phenomenon in woody plants, virtually unknown in Europe. This diversity of leaf-form is a feature that usually appears when plants are neither young nor old (Fig. 20). This also would speak in favor of a considerable "stock" of formative possibilities. Then there is the large number of hybrids, among which are species produced by crosses between members of different genera as well as between close relatives. The occurrence of such hybrids is always possible, and new ones are constantly being discovered. Some of them look completely different from the parent species, implying that there are practically no limits to the extension of the morphological spectrum.

· This, of course, is a process that goes on over the whole world; the form of a particular species of plant is very much more flexible than that of the animal, and is clearly not so strongly "individualized." Thus, at the species level, it is easier for plants to crossbreed with other species. In New Zealand, however, this phenomenon is particularly strong. Thus Peter Wardle, author of the standard work *Vegetation in New Zealand,* remarks that "progress in evolution seems to be occurring through the dominance of hybrids, in which parents of very dif-

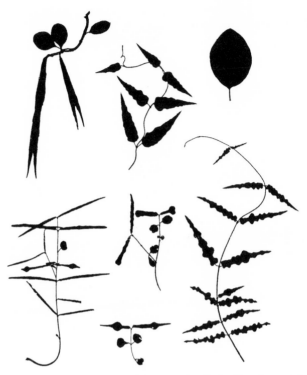

Fig. 20. Heterophyllia (variation of leaf-form) in the New Zealand liana *Parsonsia heterophylla*, a member of the dogbane family (*Apocynaceae*). Below: young plants. Above: mature forms (and the long structures of burst seed-pods). Initially the leaves are clearly unable to decide between rounded or long, thin shapes, and so produce a lively mixture of both.

ferent appearance are often involved (e.g., bushy and herbaceous species)." In this way nature is demonstrating something that humankind took up and further developed in the breeding of crop plants. Wheat, maize, potato, and many others are often the product of several "artificial" crosses (which were actually carried out in a very natural way).

What is the case for the single organism applies equally well to the large-scale ecological organism—to whole landscapes or plant communities. In this connection the well-known plant ecologist Heinrich Walter, who has firsthand experience of many countries, remarked: "Strange as it may seem, the distribution of the sub-tropical and *Nothofagus* forests cannot be explained climatically. . . . This mosaic-like alternation between two mutually exclusive forest types [i.e., the podocarpus and southern beech forests] makes it highly likely that the current plant cover is not in tune with the climate" (1968). Walter says that here the process of development is still in full swing and will probably lead to a further expansion of the beech forests at the expense of the podocarpus forests.

HUMANKIND AS DESTROYER AND AS AGENT OF FURTHER EVOLUTION

New Zealand is probably the last area of the Earth to have been brought under permanent human settlement! *Aotearoa*—as the two great islands are called in Maori—was not roused from its long sleep until the fourteenth century, when groups of Polynesian farmers, fishermen, and hunters began arriving. Nature suffered its first casualties. The moas gradually died out. Whether the Maori hunters were solely responsible for this or whether they simply finished off an animal whose time had run out anyway is still an open question. It is, nevertheless, fashionable to make human beings guilty of all extinctions that have occurred since their emergence. This attitude amounts to nothing less than a wholesale projection of modern human behavior back into times when humankind's relationship to nature was quite different. For instance, it is known for certain that the mammoths of the Northern Hemisphere were not exterminated by human activity, but died out for other, not least climatic, reasons. Perhaps both the moas and mammoths had simply outlived their time; phylogenetic ageing and death has been common in the animal kingdom from earliest times.

The real "nature shock" began—astonishingly late—with European settlement at the turn of the nineteenth century, and continues undiminished to this day. The reckless clearing and exploitation of the forests set in—among the victims the stands of kauri on North Island, for instance. But that was by no means the worst of the interference, for large expanses of primeval forest still exist, apparently about 60 percent of the original area (Wardle 1991, Smith 1988).

Much more serious was the sudden outbreak of the "age of the mammals," which came with the introduction by Europeans of certain animal companions, some intentionally, some unintentionally. Among these, two still dominate the scene: the sheep and the rat. Vast areas were given over to sheep pasture, which due to overgrazing have long been suffering ever increasing erosion—small wonder in such a rainy climate—and thus can never again be viable either as natural landscape or as farmland. The rat, which in its unmatchable adaptability resembles humans more closely than any other animal, has long ago invaded every habitat and done a thorough clearing out. The indigenous birds, unaccustomed to the presence of mammals, are all under threat (Lockley and Cusa 1980), and some species could be saved only by elaborate rescue operations in which the last survivors were transferred to remote and thankfully rat-free islands (Butler and Merton 1992). Ferrets and weasels brought in to combat the rats (and rabbits) only compound the problem, as do the many free-roaming house cats, and all three are flourishing in the wild. The introduced red deer and wild boar, whose numbers have reached uncontrollable proportions, cause considerable damage to young growth and ground vegetation in the forests, and the higher branches they cannot reach fall prey to the brushtailed phalanger ("possum"), an odd looking marsupial about the size of a marten, introduced from Australia for its fur (Brockie 1992). Although it is at the top of the list of fair game, eliminating it is literally impossible; and although large numbers are run over at night, this also has no noticeable effect on its teeming population. Dead trees, their branches stripped bare by possums, are an everyday image. Interestingly enough, in the case of the immigrant songbirds there is nothing comparable, as was mentioned earlier.

In the plant world the situation at first looks not much different. It is possible to drive for hours through regions where one sees no indigenous plants, only foreign ones—which have long since been naturalized: marguerites, wild chicory, foxglove, and an endless list of others, which could go on for pages, even whole volumes. The thickest volume of *The Flora of New Zealand* is the one dealing with immigrant species. It is 1,300 pages long, even though it has nothing on the monocotyledons—the grasses, the lily family, etc. (Webb et al. 1988). Willow bushes—not native to New Zealand—line the rivers and streams, while the general scene features Monterey pines from California, Australian eucalyptus, and rhododendron trees from Nepal. The greater part of these "neophytes" comes from Europe, with North America, Australia, and temperate Asia also strongly represented.

Nevertheless: all these newcomers add something new to the landscape, color especially. The sight of a strong red, a bright blue, of violet and orange is very exhilarating. The mind lights up through this appeal to the senses, and the mild soporific stupor induced by the dull greens and browns of the native forests and scrublands is blown away. Freestanding, nonnative trees very often cut an impressive figure in the landscape. It seems that here they can express their full "personality" in a particularly powerful way.

These might be felt to be very subjective impressions that communicate nothing of the essential nature of the phenomenon. After all, a stag also presents a noble picture, and the possum is certainly cute, but this level of experience says nothing about the ecological effects these animals have on their surroundings. By the same token, it is necessary to inquire whether the role these newcomers to the plant world of New Zealand play in the natural scheme of things is detrimental or not.

Most of them are species that *follow in the train of culture*. In their country of origin they tend to grow in places where nature has been changed in some way by human action: fallow and arable lands, meadows, cleared forest lands, the sides of roads and paths, derelict places, etc.—and in New Zealand they are found in just the same sorts of places. In keeping with their own requirements they generally also stop short of encroaching upon areas of primal wilderness, especially forest. Should they encroach, they do so only along the edges of paths. These are not merely the observations of a short-term visitor, but stem from the grand old man of botany in New Zealand, Leonard Cockayne, whose assessment is also endorsed by modern investigators (Cockayne 1921; Wilson 1976).

Naturally there are exceptions. Willows and lupines throng the wide gravel beds where rivers emerge from the mountains onto the plain, thus robbing the birds that breed there of their habitat and threatening their survival. And there are other examples. All in all, however, there is nothing comparable to the impact the introduced mammals have had.

The situation is quite the reverse, in fact. There can be no doubt that New Zealand has been enriched through the plants that came in on the heels and through the hand of humankind. They added something previously missing that nature, in its original local form, was incapable of producing. Here the influence of human beings had two sides, as indeed it does everywhere: on the one hand they drove back the purely vegetative element, but in so doing they released much formative potential in the original ecosystems, which were subsequently capable of taking up and integrating new elements. These elements were an expression of higher mental activities—of the aesthetic as well as the purely survival needs of human beings, and this, together with the fact that they introduced numerous higher mammals, adds up to a new phase of coevolution between humans and nature.

FUTURE DEVELOPMENT?

All over the world we are seeing the growth of a feeling of responsibility toward nature. In New Zealand also, increased efforts are being made to preserve and protect the natural heritage. That this partly involves calls for all "foreign elements" to be eliminated is only to be expected. This attitude, however, is itself unnatural, since nature, as has been amply demonstrated, is constantly in motion, a flowing process in which the distributions of plants and animals change even in the absence of human intervention.

The communities of commercial plants found on traditional as well as modern organic farms are largely "natural" and could in many ways be taken as an ideal model: plants from all corners of the world, nurtured by humans, have formed new communities, in association with a much larger number of wild species. Nowadays, and this goes as much for New Zealand as any other part of the world, this human husbandry can no longer be restricted to cultivated land but must also take account of so-called virgin nature.

The case of New Zealand shows just how important the preservation and protective care of the original natural landscape is and how necessary it is for it to exist on an equal footing with the cultivated landscape. *It is the source and refuge of the vegetative potential that makes New Zealand so unique.* Should these veritable reserves be virtually or completely destroyed, then the whole background of vegetative life, upon which a healthy agriculture is built and which enables it to flourish, would disperse and be lost forever.

All the more important for the future, then, is a form of agriculture that not only takes nature into account but also knows how to *think as nature thinks.* That will be the only recipe for sustainable success, and it will mean overcoming both the mistakes of colonial times and those of modern industrial agriculture, mistakes that resulted from humans paying no heed to the natural qualities of a particular country but recklessly forcing their own demands upon it.

In New Zealand's case it is certainly possible to discover in nature pointers to pathways that could be followed. The high vegetative potential that we have observed, which seems poised in a state of almost pious expectation, as if waiting to be shown the way, can and must be taken in hand, for that way lies real evolutionary progress. The way foreign flowering plants visibly thrive here and, without any ill effects, add new qualities to the original natural vegetation speaks a clear language.

With the mammals the situation looks different. Experience shows that in New Zealand the natural environment is so dominated by plant life that it cannot cope with this animal element. This is certainly true, but might not things be different if the animals in question could be so ecologically integrated that they would not merely deplete and consume, but would also contribute in their way to the preservation and regeneration of the natural vegetative vitality? This implies animals that would themselves be involved to a large extent with vegetative processes, in other words, ungulates. Not necessarily sheep, with their dry, poor quality dung, but cattle. These would of course not be kept in "monocultures," but as integral parts of a mixed agriculture, in which they would contribute to preserving and increasing the soil fertility. *Such a biodynamic method of farming, which understands the right way to combine plant and animal husbandry, might well be what New Zealand is waiting for.* This, of course, in no way implies destruction of the natural environment. Indeed, its total destruction would not only call into question the growth of a healthy agriculture but would threaten the development of human culture as a whole. Those who find this unintelligible need only ask themselves how culture comes into being at all. It is a human artifact and is not created out of nothing, but is in truth a transformation of the vegetative potential not needed for the normal growth processes of the human body. The "youthfulness" of human beings as compared to animals rests upon this store of formative potential, which in due course is used for shaping and realizing spiritual intentions, being transformed in the process: culture is secondary nature, nature heightened and enhanced.

Of course, in the creation of culture, that is, the transformation of the Earth, the formative reserves of nature are also involved. This is made clear by agriculture and horticulture, where human ingenuity and the vegetative potential of cultivated nature merge into one. *The result is an intensification of both—in other words, culture.* This also means, however, that both are equally essential and that the disappearance of either would jeopardize the maintenance of culture. Here the danger signs are all too apparent.

The Signature of the Great Rift Valleys

PARALLELS IN THE HISTORY OF EARTH AND HUMANITY

IT IS ALWAYS SURPRISING when landscapes far apart both in location and character present us with a very similar profile—steep mountain ridges, say, with contours so straight they might have been drawn by a ruler. Letting one's gaze travel, for instance, along the sheer wall of the Baikal Gorge, at the point where it runs southwest abutting the southern slopes of the Sajan Mountains, one is forcibly struck by the resemblance it bears to sections of the East African Rift Valley in the region of Lake Manyara. This impression is considerably reinforced by

Fig. I. The wall of the East African Rift Valley at Lake Manyara in northern Tanzania

reconstructing in imagination the lakebed that once existed where now the river Irkut flows through the alluvial plain. A similar picture emerges at a number of places in the valley of the Jordan, and even along the Rhine valley where it meets the eastern edge of the Vosges, albeit in a less dramatic form.

What most of these great valleys have in common is that they are part of a system of dilation zones spanning the whole Earth. These are places where viscous molten rock under enormous pressure is surging toward the surface, only to be repelled to right and left under the Earth's crust. As a result, the crust is subject to a strong tearing motion. Such dilation zones run down the centers of all the Earth's ocean beds, where they take the form of ridges caused by crystallization of the molten material constantly flowing out of the rift. Where they occur on land, however, they gouge out great gorges along the rifts, or produce the occasional volcanic eruption. In contrast to the dilation zones stand the compression zones, areas where slowly flowing streams of viscous rock in the Earth's mantle, together with the tectonic plates drifting along on top of them, collide. As a result of such collisions the continental plates buckle and fold, forming mountains in the process, while the oceanic plates are pushed down into the depths.

It would thus appear that rift valleys are the polar opposites of

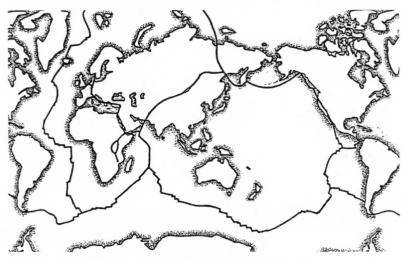

Fig. 2. he Earth's dilation zones, seabed ridges, and continental rift valleys.

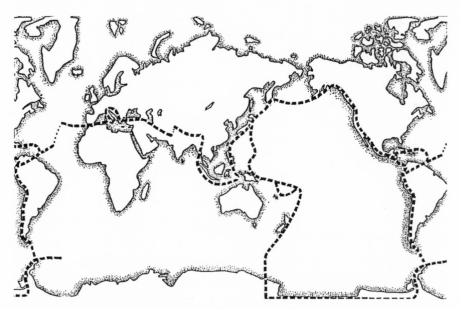

Fig. 3. Zones of compression and subduction. On land they are the locations of the most recently formed high mountain ranges.

mountain ranges, both in their topography and in their mode of formation; they are, as it were, "negative mountains," and vice versa. This impression of polarity or complementary opposition is reinforced when we focus on the related processes of earthquakes and volcanic activity. The compression (mountain formation) zones are the areas where silicon-rich eruptions and violent earthquakes occur frequently (all around the Pacific and along the mountains of southern and southeastern Europe and the Near East from Sicily to Iran). In the continental dilation (rift valley) zones, earthquakes and volcanic eruptions are not unknown, but tend to be milder. Often all that happens is a local buildup of subterranean heat. The magma sources feeding such basaltic hot spots lie in deep layers of the Earth's mantle not subject to drift. Evidence of such an oceanic hot spot—or, more particularly, of its stationary source—is found in the chain of extinct volcanoes stretching underwater for thousands of miles to the north and west of Hawaii. The further from the source, the older they are, for each in turn has been carried along by the lateral movements of the oceanic

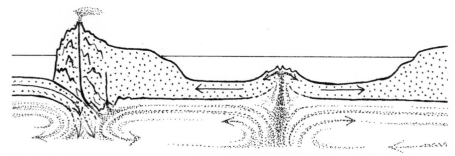

Fig. 4. Schematic diagram of the relationship between compression (left) and dilation zones (right), roughly corresponding to that between South America and Africa. The upward convection movements produce ridges in the ocean floor under which the molten rock surges upward, only to be forced to either side, taking the plates floating on top of it along with it. When two plates collide (left) the oceanic plate involved will be pushed under the heavier continental plate, which in the process undergoes folding with the consequent formation of high mountain ranges. Large bodies of magma force their way through cracks and then actively melt a pathway upward to form volcanoes on top of the mountains. Earthquakes occur as a result of mechanical disturbances where two plates are rubbing along each other. The diagram also makes clea why deep-sea gorges form in the subduction zones, where an oceanic plate is being pushed downward.

plates, while the hot spot, with its deep source, has held its ground and there continued to cause uplift. The lava that emerges when an actual eruption occurs at such a hot spot has a chemical composition very different from that of the compression zones. Whereas the latter is most likely to be acidic in nature, the lava flow from a hot spot volcano will usually be found to be alkaline (Schmutz 1996).

Polarity appears also in the geographical distribution of compression and dilation zones. While the former are found almost exclusively along coastlines or inland (although in the case of the Alps or the Himalayas they also follow what used to be coastlines), the latter belong primarily to the depths of the oceans, as the term midoceanic ridge suggests.

Where two plates meet—be it along a line of compression or dilation—a third form of contact can take place. Instead of pulling apart or colliding, the plates can rub along each other. Such horizontal shifts, in other words earthquakes, are a feature mostly of compression zones, well known to occur, for instance, in Turkey or along the infamous San Andreas Fault that runs through the suburbs of San Francisco. The same phenomenon appears, this time in a dilation zone, in the

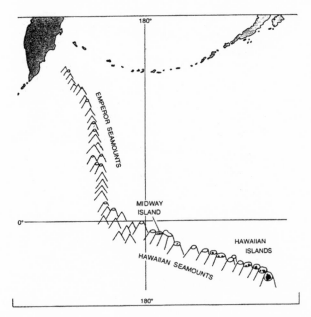

Fig. 5. A chain of extinct, submarine volcanic craters marks the path of the Pacific plate over a "hot spot" during the last 70 million years. Its present location is Hawaii. (After Heezen and MacGregor 1973.)

Near Eastern section of the Great Rift Valley, where the side east of the Jordan is moving northward.

In the global pattern of compression and dilation zones we can discern the edges of two interlocking tetrahedrons, the "compression tetrahedron" forming one pole and the "dilation tetrahedron" the other.[1] The sparsely documented fault line running from Pamir through Lake Baikal fits into this pattern. *The African-Middle Eastern Great Rift Valley, however, does not.* This formation, together with its northern extension, with which we will be concerned in what follows, does not tally with the tetrahedral structure. It would seem to relate to some other pattern. It is also exceptional in being the only major system on land in which dilation processes predominate.

---

1. This theory traces back to a reference made by Rudolf Steiner to the effect that the Earth has had the form of a tetrahedron imprinted upon it, whose points coincide with particular geographical locations. Initially quite independently of this the Swiss geologist H. U. Schmutz noticed that the systems of compression (subduction) and dilation zones are mirror images of each other (1986). Transferring these to the spherical surface of the Earth, he found that the sets of lines together described a tetrahedron (thus preserving the mirroring effect).

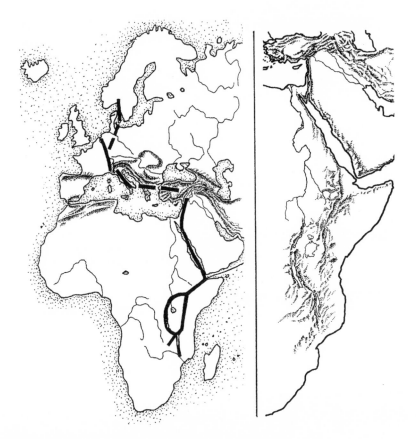

Fig 6. The course of the African–Middle Eastern–European rift valley system. (Combined from S. Müller 1984 and H. U. Schmutz 1986.)

THE AFRICAN-MIDDLE EASTERN-EUROPEAN RIFT SYSTEM

The Rift Valley brings a strong element of scenic drama to the monotony of the flat, if slightly undulating, African plains. The charm of the landscape suddenly increases. It is full of strikingly individual forms, such as giant mountains, standing alone, each with its own "personality." Starting on the Zambezi at the northern border of Southern Africa the Rift Valley branches in two through Central and Eastern Africa, then joins again and continues through the Red Sea to become the

Fig. 7. The view from Oldeani in northern Tanzania of Kilimanjaro (5895 m) and Mount Meru (4565 m) through a strong telescope from a distance of between 140 and 200 km.

Dead Sea Depression and the Jordan Valley. From there it runs northward between the Galilean Plateau and the Golan Heights, and then along the inland edge of Lebanon, ending up in the Turkish part of Syria (Hatay, formerly Antioch).

To take a walking tour starting from the northeastern corner of the Mediterranean and heading south over the Amanus Mountains (Nur-Daglari) is a very instructive experience, especially in springtime. At first the vegetation is typically Mediterranean, with olive groves and citrus plantations, gradually changing, as we climb through stands of cedar, fir, and then beech into the central-European vegetation of the uplands (cowslips, violets, etc.). Then, as we descend, the landscape all at once takes on the features of a genuinely Palestinian gorge. When we look back, the western wall we have just descended presents a truly imposing spectacle, whereas the gorge's eastern flank takes the form of a chain of low hills (the Jebel el Akrad, part of Syria). Here in the dry steppe we encounter plants common in Galilee and Samaria: scarlet crowfoot, Mount Carmel orchid, and many others. There are also large concentrations of migrating birds that follow the Rift Valley all the way up from Israel: vast flocks of storks rest on the sheep meadows; cohorts of pelicans, lesser spotted eagles, and steppe buzzards by the hundreds continually cross the sky, heading north.

In that direction, however, the end of the valley is already in sight. The east-west walls of the Taurus Mountains of southern Anatolia seal the area off. The migrating flocks react directly to this obstacle. The buzzards turn to the northeast, following the mountains of the "Anatolian diagonal" toward the Armenian Plateau, the Caucasus, and beyond, while the pelicans, storks, and eagles head northwest, making for

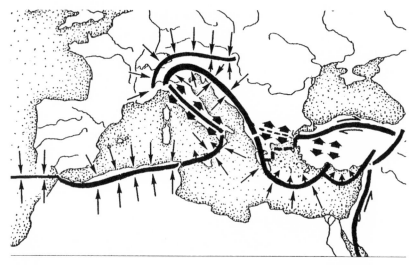

Fig. 8. The northern edge of the African plate, showing the "Adriatic Spur" and the Aegean and Anatolian microplates. The thin arrows mark the compression zones, the regions where collision, subduction, and folding are occurring. Thick arrows denote dilation zones, the regions where plates are drifting apart, and small arrows show zones of horizontal shift. (After Müller 1984, amended.)

destinations in Asia Minor and southeastern and eastern Europe. The end of the gorge does not, of course, signify the end of the fault line. It carries on as a zone of dilation and/or horizontal shift associated with the Anatolian plate, and now runs almost due east-west parallel to the compression zone where the Eurasian and African-Arabian plates meet. Shifts do occur periodically along this line (Fig. 8) and are likely to figure high on the seismic scale. During a single earthquake it is not unknown for the areas lying north of this line to have shifted several meters to the east (Pavoni 1969). Via the "Aegean microplate" the line continues on toward the Apennine Peninsula, where, according to the latest research, it follows the "Adriatic Spur" of the African plate by going *underneath* the Apennines.[2] (In all probability the 1997 earthquakes around Assisi were connected with this.)

Slightly further west the "so-called 'Central European rift system,' which stretches all the way from the western Alps to the North Sea,"

---

2. The African continental plate is drifting northward with a counterclockwise rotation, which causes ruptures of many kinds in southern Europe where it collides with the Eurasian plate. The intrusion of the part of Africa called the "Adriatic Spur" led to the folding process that formed the Alps. To this northern spur of the African plate belong "Sicily, the whole Apennine Peninsula, the Po Basin, the southern Alps, as well as the Adriatic with the western coasts of [former] Yugoslavia, Greece and Albania" (Müller 1984).

joins on. "The extreme thinness of the lithosphere in this system indicates processes going on in the depths that might one day cause the European continental plate to break asunder at this weak point. This rift zone, at present only mildly active,[3] reaches from the western North Sea down through the Netherlands, the Rhine delta, along the valley of the Upper Rhine, under the French-Swiss Juras into western Switzerland and then finally southwest under the western French Alps" (Müller 1984). A (currently) passive side branch runs through central and northern Germany and then along the west coast of Sweden as far as the Oslo Fjord.

PARALLELS IN THE DEVELOPMENT OF THE EARTH AND HUMANITY

The system of rift valleys we have been considering is, as an inland zone of dilation, unique; but it is also very striking in its constant close connection to human evolution and cultural development over all the millions of years that have elapsed from humankind's beginnings to the present day. Starting from the south, it is as if it had been carved in the Earth as an axis showing the northward course of this development. Accordingly, finds of the earliest humanoid remains (*Australopithecus*), three to four million years old (some possibly older), were all made either within or near the East African Rift Valley. The famous footprints found near Laetoli in what is now northern Tanzania are of a similar age. They were made by three prehominids with upright gait, an adult and two children, in a long since solidified carpet of volcanic ash (Leakey 1979; Leakey and Hay 1979).

Then about two and a half million years ago the first genuine hominids, *Homo habilis* ("the skillful"), appeared on the scene (Schad 1985). Stone tools fashioned by their hands have been found in various places all over East Africa. Still nearer to the human form as it is

---

3. A keen reminder that this zone is not quiescent came with the earthquake on Easter Monday, 1992. Its epicenter lay right in the fault line, just like that of numerous earlier tremors (e.g., the catastrophe that hit Basel in 1356).

now was *Homo erectus,* who appeared two million years ago (Schad 1985). The earliest evidence of humankind outside Africa was found near Ubeidya, slightly south of the Sea of Galilee. This was in the form of skull fragments judged to be from *Homo erectus* individuals of a similar age to those known from finds made at Lake Turkana in Kenya and in the Olduvai Gorge in Tanzania. The Jordan Valley is the most northerly extension and at the same time the youngest part of the East African Rift Valley system. In East Africa the Rift Valley dates from the early Tertiary period, whereas the formation of the Red Sea, the Araba Basin, and the Jordan Valley took place in mid-Tertiary times (the Miocene). A general lowering of ground level set in during the late Tertiary (Pliocene); it increased in pace during the Ice Age (Pleistocene) and is still in progress. The Jordan is, as geographers are fond of expressing it, "the youngest river on Earth." Hominids thus lived in Ubeidya at a time when the Jordan Valley looked very different.

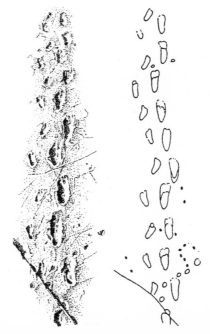

Fig. 9. The famous footprints of an adult and two children (one of whom was stepping in the adult's footprints) near Laetoli in Tanzania—according to current reckoning 3.7 million years old. (Drawn from a photograph, after H. M. Muller 1987.)

Through countless generations they were constant witnesses of the opening of this great cleft with its river, and also experienced the massive eruptions on the Golan Heights when great streams of molten basalt poured into the upper Jordan Valley north of the Sea of Galilee. From the beginning to the middle of the Pleistocene, prehistoric humans were party to these momentous events. "Here the Earth's history and human history have assimilated each other" (Schad 1996).

The Jordan Valley and the Arabia-Dead Sea Basin together form a region that is ecologically unique. Here the Rift Valley provides a meeting place for living communities from north, south, east, and west. While animals and plants from northern temperate zones migrated through the hills that flank the Jordan, eventually making their way into tropical Africa (e.g., ibex, juniper), elements of tropical Africa penetrated northward through the shelter of the valley (e.g., antelopes, ostriches, acacia, doum palm, papyrus), pointing the way, as it were, for early humans. From the east came flora from the desert and steppe regions of central Asia; and in the carpets of red anemones, yellow-eyed daisies, and purple sage that cover the landscape in spring can be seen the long-standing influence of the wet Mediterranean-Atlantic climate of the west, which also provided the whole basis for the agricultural development of the region (Suchantke 1996).

It was along these same pathways that the eventual dispersal of *Homo erectus* took place. The Jordan Valley is the only overland gateway out of Africa, and from here this early form of humanity spread far into Europe and over the whole of temperate and tropical Asia (e.g., Peking man, Java man). Although they are by no means confined to them, the finds of *Homo erectus* in Europe also cluster along the rift valleys. Their wanderings led them into the valley of the Rhone, whence they "discovered the Burgundian gap between the Swiss Juras and the Vosges, followed the plateau of the Upper Rhine, a further rift valley landscape, continued through Wetterau, the Kassel basin, the valley of the Leine in Lower Saxony and into northern Germany and Denmark. Upon turning 'back' they would mostly keep to the same thoroughfare. Thus *Homo erectus* (and *Homo praesapiens*) had no settled existence. Evidence has been found of resting places that were repeatedly visited, and always at the same time of year, for example near Nizza (Terra Amata). Like migratory birds, early humankind moved north in summer and south in winter; roaming vast distances" (personal communication, W. Schâd).[4] The next in line, *Homo sapiens*, likewise emerge out of Africa. The earliest evidence for them stems from the last ice age but one (the Mindel glaciation),[5] and then, during the last one (the Wurm glaciation), they managed the step to the New World, and with it established their presence over the whole Earth.

Some considerable time later, but still long before the advent of written records, this same area saw the occurrence of the key events that formed the basis for the flowering of the cultures of the Middle East, North Africa, and Europe. About 9,000 B.C. we have, in Jericho (the world's oldest urban settlement), the earliest known use of cultivated wheat. Before this the inhabitants of Jericho had simply gathered wild wheat, which still grows in the form of steppe grass in "natural cornfields" all over the Near East. Wheat was soon followed by oats and barley, and in all likelihood by the cultivation of olive trees. The domestication of sheep and goats also belongs to this area, if borders are not taken too strictly and further regions of the Fertile Crescent are included in the picture (Bar-Yosef and Kislev 1986; Clutton-Brock 1978; Harlan 1977; Zohary 1969; Zohary and Hopf 1993).

The central significance of this area for the whole Earth was certainly recognized in antiquity (Suchantke 1996). It was built up through the fact that the most important strands of world culture from prehistoric and historic times converged and intertwined here:

---

4. Here, of course, it may be objected that the clustering of finds along the rift valleys could simply be due to the fact their geological layers remain "fresher" than heavily eroded plateaus or plains that have been filled up with sediment. "On the other hand, the clustering cannot be explained on this basis alone. The rift valleys of Africa, the Middle East, and Europe are genuine migratory pathways. They have remained such right into historical times" (personal communication, Schad).

5. This early form is normally designated as *Homo sapiens anteneanderthalensis*, the direct ancestor of Pre-Neanderthal and Neanderthal humans and the modern *Homo sapiens sapiens*, whose earliest known remains (from Florisbad, South Africa) have been dated at 170,000 B.C. (personal communication, W. Schad).

the cultural strands of Mesopotamia, Persia, and ultimately India, of Egypt and Greece, and, in the middle, the Hebrew culture. Finally, the ministry of Christ at this focal point of the Earth's and humanity's history places all other developments in the region, whether of nature or culture, into a new perspective.

The further westward course of the continental dilation zone from Greece and the Aegean via Italy to central and western Europe marks, again with astonishing exactitude, *the axis along which developed those cultures which have spearheaded the transformation of human consciousness from antiquity into modern times,* and which are now expanding over the whole Earth. Parallel geological and cultural development on this sort of scale cannot be mere coincidence. It shows that the developmental history of Earth and humanity, of nature and culture, belong together, and that both spheres are part of one process.

OTHER RIFT-VALLEY SYSTEMS AND THE BAIKAL REGION

When we turn our attention to other continental rift valleys, what presents itself? In Africa, mostly in Guinea, but also in the north of South Africa, there are a number of smaller-scale fault systems that "have become inactive" (Schmutz 1996). In their vicinity have been significant cultures, which have long ago died out. We know virtually nothing about the ruins of Zimbabwe in the south of Africa, but we are by no means in the dark about the West African cultures of Ife and Benin, which persisted into modern times. They and their neighbors the Yoruba were the only black peoples to develop an urban culture—Ibadan is one of their towns. Above all, the artistic standard of their famous bronze sculptures is unique within the whole of Black Africa, and represents the pinnacle of its artistic expression.

In contrast, the Baikal Gorge in southern Siberia is definitely not inactive. Although it is not a recent formation and may reach back into Mesozoic times (Weber 1992), it is still in movement. This is shown both by the frequency of earthquakes and by the fact that the gorge is still getting wider (2 centimeters per year) and deeper: it is all too common for patches of ground to cave in during earthquakes. On New Year's night of 1861–62 hundreds were killed when an area of 200 square kilometers to the north of the Selenga delta sank below water level (marked as S. P.—Saliv Proval—in Fig. 10).

In its time the area has been subject to many and varied cultural currents, but they have always been streamlets rather than rivers. Mongolian shamanism has been shown to contain Zoroastrian elements, and even influences from the Mithras cult of ancient Rome (Eliade 1964). The route across the steppes from the Caspian and Black Seas has seen much coming and going over the centuries. A further revitalizing influence, increasingly so in recent times, has been that of Buddhism. The island of Olchon in Lake Baikal has become a sacred place for pilgrims from Japan, Korea, and Mongolia.

Modern influences are detectable chiefly in nearby Irkutsk. With its varied cultural life, it has a very lively air about it, especially when

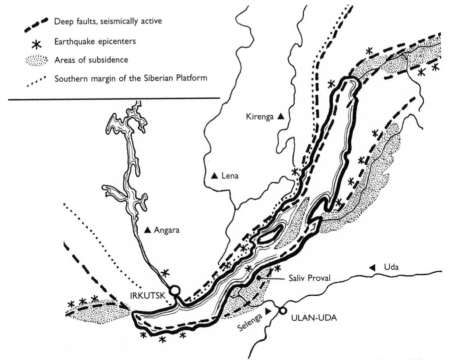

Fig. 10. The Lake Baikal region, showing fault lines and areas of seismic activity and subsidence. (After B. Weber 1992.)

compared with the gray gloom of most Russian towns. The town may hold much promise for the future because of having been used for centuries as a place of exile for members of spiritual elites—think, for instance, of the cultural and social achievements of the Decembrists (cf. Sutherland 1990; Sergejew 1986).

Exact prognoses cannot be made, of course, but whatever is "in the air" in the way of tendencies that could have a hand in shaping the future can be observed there. Moscow and European Russia are far removed, and what came, and still comes, from there has little to offer the people of this well-nigh Far Eastern outpost. Tendencies toward independence from Russia, while present, are so far not very strong, but an increase in chaos and poverty in this vast region would certainly strengthen them. Already there are distinct signs of a leaning in another direction—toward the much closer Pacific region and the economic powers of Japan and Korea. The latter especially is making inroads everywhere with "developmental aid." The more the techno-logical power (and perhaps also the spiritual and cultural influence) of these countries increases, the less Siberia will be able to resist it.

Thus the still somewhat vague possibility arises of a "Pacific block," stretching from Siberia to California, taking shape in the future. Japan and California are currently, and are likely to remain, the world lead-ers in high tech, and in both of them the Far West and the Far East are combined. This is particularly true of California. In relation to the rest of North America it lies "across the great divide," in a region where the Pacific and North American plates "rub shoulders." As a result of this contact certain areas of the West Coast have become separate "microplates," and in some places (called terranes) the Pacific plate has even been lifted out of the sea and incorporated into mountain-building processes. Although there are no terranes in California itself, they have occurred further north, in Alaska, and along the coast of South America.[6]

---

6. It is well known that the San Andreas fault is of a different and more complex nature than the rift systems so far mentioned. It represents the contact line of both a compression zone (the Rocky Mountains) and a dilation zone (an oceanic ridge under the Gulf of California). The net result is that the Californian microplate is shifting northward. What we have on land, therefore, is not a dilation zone—this is proved by the existence of the western moun-tain ranges, which are the result of contact between plates and consequent tearing. As already

California is also the place where the voice of the prophets of the New Age—those trying to forge a synthesis between Eastern and Western forms of consciousness—has been heard the loudest. Up to now this has largely entailed mixing elaborate philosophical cocktails in which the most abstract conceptual constructs of modern science are stirred in with images culled from mythic and cosmological sources. With such incompatible ingredients, belonging as they do to radically different layers of consciousness, it is no surprise that the results tend to be either unpalatable or too sweet.

A mental attitude that always looks for depth, clarity, and inner coherence in thinking finds it easy to denigrate such New Age writings as a conceptual jumble, but that may well be to ignore the fact that behind them lie the first stirrings of a genuine urge to create a comprehensive global consciousness. Here, of course, "global" should not be taken merely in its spatial sense, but much more in its temporal, evolutionary sense—meaning that human beings have now reached a time when they should be able to have simultaneously at their mental fingertips all forms of consciousness that humanity has gone through in the course of cultural development, and to understand them all as part of one great Whole.

To understand what this implies, a comparative look at the developmental steps involved in the life of an individual human being may be instructive. The abilities for learning, spontaneity, and imaginative play that feature large in early childhood should not be jettisoned during the process of growing up. They should rather be "kept on board," if the adult is not to end up as someone who cannot think or act on his or her own initiative. Of course, for adults to preserve these abilities acquired (and nurtured) in childhood does not mean that they will be childish; the ability for detached reflection upon oneself and others that comes later *transforms* the heritage of childhood so that *it*

---

mentioned, they consist of parts of both the major plates: certain coastal ranges of the North (and South) American cordilleras are uplifted and folded ocean floor. The principle that mountains form a barrier between different bioregions applies here too: California, lying, as it does, on the other side of the mountains, does not belong to the rest of North America, but to the realm of the Pacific. To repeat, it is Far West and Far East at one and the same time.

can now be used consciously. By the same token it would not be "atavism," not a regression to earlier (unreflecting and uncontrolled) forms of consciousness, if our slumbering capacities for visionary and other modes of inner experience leading to concrete encounters with spiritual realities were awakened. In principle this is possible because we carry these capacities within us from the childhood days of humankind's development—we have simply forgotten them.

More precisely, the point is that such a synthesis of types of consciousness *could* be—or could become—a path of spiritual development, not of some transpersonally united collective, but of the individual self. Whether the Far Eastern component in this cultural fermentation process will countenance that remains to be seen.

THE SHAPING OF THE EARTH

What has all this got to do with rift valleys? Perhaps more than might appear at first glance. The rifts are like great flesh wounds in the body of the Earth, and in some places they are very deep indeed: Lake Tanganyika in the East African Rift Valley is 1,435 meters deep and at its deepest point descends to 662 meters below sea level; Lake Baikal is 1,637 meters deep and reaches to 1,180 meters below sea level. The record inland depth is that of the Dead Sea; although the lake floor is "only" 793 meters below sea level, the actual rift is several kilometers deep at this point, having been filled up with rock debris.

According to the conventional view, volcanic processes going on under the Earth's crust are ultimately what set the terms for life on its surface. Thermal processes in the Earth's mantle create large circulatory movements in the molten rock. The plates of the crust swimming on top with their continental cargo are carried along and thus in the course of time the continents are "shipped" through various climatic zones. The long-term transformation of European flora and fauna from tropical to temperate was not due so much to climatic change (ice ages notwithstanding) as to Europe's drifting, having been nudged northward by Africa.

Mountains are formed when differing, spatially distinct tectonic plates collide and merge, as, for instance, in the Alps, which are made up of European and originally African parts. (To take a day's hike from the central into the southern Alps is to cross from one continent to the other, and at certain spots it is possible literally to stand with one foot in Europe and one in what was Africa.)

All over the Earth mountains stand as a check to the dispersal of animals and plants. As climatic watersheds they form barriers between different ecological regions. Anyone who has ever crossed the Alps—or the Scandinavian mountains, the Andes, the Himalayas—knows from personal experience how this transition means not only a striking change in the natural environment, but also a corresponding alteration in one's inward responses to it.

Rift valley systems present exactly the opposite picture. Here originally uniform plates split and begin drifting apart. Instead of physical matter accumulating and being piled up, as in the case of mountains, it is diminished in bulk and thinned out right down to considerable depths. In a clear countermovement, living processes stream in from all sides in a great intensification or perhaps even "potentiation" of generative activity. The incredible concentration of animals in the region around the Great Rift Valley of East Africa has no equal on Earth. The ecologically identical savanna areas of South America or Australia have nothing to compare with it. The intensification of living process in the region of the Dead Sea and the Jordan River has already been indicated above in the description of its animal and plant life and bird migrations. The same thing is clearly detectable also in the Rhine-Rhone valley, that migratory pathway for Mediterranean animals and plants.

That this intensification of living process is a feature belonging particularly to zones where the physical world has been thinned down is shown by the fact that in the depths of such zones life can take hold as it does nowhere else. It is true that life exists in the oceans at all levels, even in the very deepest waters, where creatures have been found that live off the constant shower of organic debris percolating down from the upper layers. Their populations, however, tend to be somewhat thin. In the rifts that split the oceanic ridges in the ocean floor the picture is quite different. Here completely autarchic communities, independent of the light-filled surface waters, have recently been discovered. They are made up of shellfish, worms, crabs, and shrimps of giant proportions (Edmond and von Damm 1983). They live at places where there are hot springs containing metallic salts. The decisive factor enabling life to exist in these zones is the presence of autotrophic bacteria that can assimilate the constantly upwelling hydrogen sulphite, a substance deadly poisonous to most living things. It was a great surprise when a Russian-American research team recently discovered very similar communities on the floor of Lake Baikal. Although, in line with the freshwater conditions of the lake they contain a completely different combination of species, the basis of their existence is the same as in the oceanic rift systems—bacteria that assimilate hydrogen sulphite (Weber 1992).

The rift valleys thus represent lines along which the physical is "attenuated" and life processes "intensified." This applies most strongly to the great formation that stretches from East Africa through the Middle East and into Europe, the only rift valley of such intercontinental dimensions on land. Its special significance is further emphasized by the fact that at a point on the Earth where there was an impenetrable east-west barrier of mountains, it forged a breach, creating a north-south pathway. Thus a two-way tide of intensified life was able to flow between Africa and the north. Seen in this light the Great Rift Valley must be regarded as one of the main arteries for the natural and cultural lifeblood of the Earth. This is a geological phenomenon, whose full reality is to act as a vehicle for a process of cosmic proportions through which the incarnation of humanity was prepared. When it came, this event thus took place in a context that provided the requisite abundance of vibrant life to enable the new inhabitants to begin a great work of transformation—the making of culture within nature.

References

Adams, N. M. 1994. *Seaweeds of New Zealand.* Christchurch.

Agnew, A. D. Q. 1974. *Upland Kenya wild flowers.* Oxford.

Amadon, D. 1973. Birds of the Congo and Amazon forests: a comparison. In B. J. Meggers, E. S. Ayensu, and W. D. Duckworth, eds., *Tropical forest ecosystems in Africa and South America: a comparative review,* 267–277. Washington.

Anderson, A. B., ed. 1990. *Alternatives to deforestation—steps toward sustainable use of the Amazon rain forests.* Columbia University, New York/Oxford.

Aubréville, A. 1949. *Climats, forêts et désertification de l'Afrique tropicale.* Soc. Ed. Géo. Mar. Colon. Paris.

Bacchus, R. 1994. Geological origins of New Zealand. Anthroposophy at Work, *Journal of the Anthroposophical Soc. in New Zealand* 1:34–37.

Bakker, R. T. 1975. Experimental and fossil evidence of the evolution of tetrapod bioenergetics. In D. Gates and R. Schmerl, eds., *Perspectives in Biophysical Ecology.* Berlin, Heidelberg, New York.

Bar-Yosef, O., and M. E. Kislev. 1986. Earliest domesticated barley in the Jordan Valley. *National Geographic Research* 2:257.

Bauchop, T., and R. W. Martucci. 1968. Ruminant-like digestion of the langur monkey. *Science* 161:698–699.

Beck, L. 1974. Okosystem tropischer Regenwald. *Bild der Wissenschaft* 11, Nr. 10:42–48.

Bigalke, R. C. 1974. Ungulate behaviour and management, with special reference to husbandry of wild ungulates on South American ranches. In V. Geist and F. Walther, eds., *The behaviour of ungulates and its relation to management.* IUCN Publications new series No. 24:830–852.

Bockemühl, J. 1966. Bildebewegungen im Laubblattbereich höherer Pflanzen. In W. Schad, ed., *Goetheanische Naturwissenschaft,* Vol. 2, *Botanik.* Stuttgart.

———. 1995. Morphic movements in the vegetative leaves of higher plants. In J. Bockemühl and A. Suchantke, *The metamorphosis of plants.* Novalis Press, Cape Town.

Bourgeron, P. S. 1983. Spatial aspects of vegetation structures. In F. B. Golley, ed., *Tropical rain forest ecosystems—structure and function*. Ecosystems of the World, 14 A. Amsterdam.

Bourlière, F. 1973. A comparative ecology of rain forest mammals in Africa and tropical America. In B. J. Meggers, E. S. Ayensu, and W. D. Duckworth, eds., *Tropical forest ecosystems in Africa and South America: a comparative review*, 267–277. Washington.

Braun, A. 1851. *Betrachtungen über die Erscheinung der Verjüngung in der Natur, insbesondere in der Lebens- and Bildungsgeschichte der Pflanzen*. Leipzig.

Britton, H., and P. J. G. Ripley. 1963. *A simple history of East Africa*. London, Glasgow.

Brockie, R. 1992. *A living New Zealand forest*. Auckland.

Brown, L. 1959. *The mystery of the flamingos*. London.

———. 1966. *Afrika*. Munich, Zurich.

———. 1970. *African birds of prey*. London, Glasgow.

Brownsey, P. J., and J. C. Smith-Dodsworth. 1989. *New Zealand ferns and allied plants*. Auckland.

Brücher, H. 1968. Südamerika as Herkunftsraum von Nutzpflanzen. In E. J. Fitkau e. a., ed., *Biogeography and ecology in South America*, 251–301. The Hague.

———. 1977. *Tropische Nutzpflanzen. Ursprung, Evolution und Domestikation*. Berlin/Heidelberg/New York.

Buechner, H. K., and H. C. Dawkins. 1961. Vegetation change induced by elephants and fire in Murchinson Falls National Park, Uganda. *Ecology* 42:752–766.

Bünning, E. 1947. *In den Wäldern Nord-Sumatras*. Bonn.

———. 1956. *Der tropische Regenwald*. Berlin, Göttingen, Heidelberg.

Burton, J. 1973. *Animals of the African year—the ecology of East Africa*. London.

Butler, D., and D. Merton. 1992. *The black robin—saving the world's most endangered bird*. Oxford.

Butzer, K. W. 1959. Studien zum vor- und frühgeschichtlichen Landschaftswandel der Sahara, III. Die Naturlandschaft Ägyptens während der Vorgeschicte und der Dynastischen Zeit. Abh. Akad. Wis. Litr. Mainz Math.-naturw. Kl., 43–122.

Campbell, B.. 1985. *Ökologie des Menscen.* Munich.

Carcasson, R. H. 1964. A preliminary survey of the zoogeography of African butterflies. *East Afr. Wildlife Journ.* 2:122–157.

Carpenter, G. H. D. 1925. *A naturalist in East Africa.* London.

Cemmick, D., and D. Veitch. 1987. *Kakapo country.* Auckland.

Chagnon, N. A. 1966. *Yanomamö—the fierce people.* New York.

Chambers, S. 1989. *Birds of New Zealand—locality guide.* Hamilton, NZ.

Clutton-Brock, J. 1978. Early domestication and the ungulate fauna of the Levant during the Prepottery Neolithic Period. In W. C. Brice, ed., *The environmental history of the Near and Middle East since the last ice age.* London.

Cockayne, L. 1921. The vegetation of New Zealand. In A. Engler and O. Drude, eds., *Die Vegetation der Erde,* Vol. 14. Leipzig. Reprint 1982.

Coe, M. J. 1967. *The ecology of the alpine zones of Mount Kenya.* The Hague.

Cooper, R. A. 1989. New Zealand tectonostratigraphic terranes and panbiogeography. *New Zealand Journ. Zool.* 16/4:699–712.

Crosby, A. W. 1988. *Ecological imperialism: the biological expansion of Europe, 900–1900.* 5th ed. Cambridge.

d'Abrera, B. 1990. *Butterflies of the Australian region* 2nd ed. Melbourne.

Daghlian, C. P. 1982. A review of the fossil record of monocotyledons. *Bot. Review* 47:517–555.

Darlington, P. J., Jr. 1969. *Biogeography of the southern end of the world.* 2nd ed. Harvard Univ. Press, Cambridge, Mass.

Dawson, J. 1988. *Forest vines to snow tussocks: the story of New Zealand plants.* Wellington.

Delamare-Deboutteville, C. 1951. Microfaune du sol des pays tempérés et tropicaux. Vie et Milieu, Suppl. No. 1. Paris.

Dorst, J., and P. Dandelot. 1972. *Säugetiere Afrikas.* Berlin, Hamburg.

Durrell, G. 1954. *The bafut beagles.* Penguin Books.

———. 1957. *The overloaded ark.* Penguin Books.

Edmond, J. M., and K. von Damm. 1983. Heiße Quellen am Grund der Ozeane. In *Ozeane und Kontinente, Spektrum d. Wissenschaft, Reihe: Verständliche Forschung,* 216–229. Heidelberg.

Egger, K. 1975. Traditioneller Landbau in Tansania—Modell ökologischer Ordnung? *Scheidewege* 5, Nr. 2.

———— and B. Glaeser. 1975. Politische Ökologie der Usambara-Berge in Tansania. Kübel-Stiftung Bendsheim.

Eibl-Eibesfeldt, I. 1984. *Die Biologie des menschlichen Verhaltens*. Munich.

Eiten, G. 1972. The cerrado vegetation of Brazil. *Botanicl Review* 38:201–341.

————. 1974. An outline of the vegetation of South America. *Symp*. 5th Congr. Intern. Primat. Soc. Tokyo.

Eliade, M. 1964. *Shamanism: archaic techniques of ecstasy*. Trans. W. R. Trask. Bollingen Series 76. Princeton Univ. Press, Princeton.

Engler, A. 1910, 1976. Die Pflanzenwelt Afrikas, Vol. 1. In A. Engler and O. Drude, eds., *Die Vegetation der Erde*. Leipzig (1910). Reprint Vaduz1 (1976).

Etchécopar, R. D., and F. Hué. 1964. *Les oiseaux du Nord de l'Afrique*. Paris.

Falla, R. A., R. B. Sibson, and E. G. Turbott. 1980. *Collins guide to the birds of New Zealand and outlying islands*. Auckland/London.

Fisher, F. J. 1965. *The alpine Ranunculi of New Zealand*. Botany Division, Dept. of Scientific and Industrial Research, Wellington.

Fosberg, F. R. 1973. Temperate zone influences on tropical forest land use: a plea for sanity. In B. J. Meggers, E. S. Ayensu, and W. D. Duckworth, eds., *Tropical forest ecosystems in Africa and South America: a comparative review*, 345–350. Washington.

Franquemont, C., et al. 1990. *The ethnobotany of Chinchero, an Andean community in southern Peru*. Fieldiana, Botany, New Series No. 24. Field Museum of Natural History. Chicago.

Frieling, H. 1937. *Die Stimme der Landschaft*. Munich, Berlin.

Gade, D. W. 1975. *Plants, man and the land in the Vilcanota Valley of Peru*. The Hague.

Galloway, D. 1985. *Flora of New Zealand, lichens*. Wellington.

————. 1992. *Checklist of New Zealand lichens*. Wellington.

Gardi, R. 1970. *Sahara*. Bern.

George, U. 1969. Über das Tränken der Jungen und andere Lebensäußerungen des Senegal-Flughuhns, Pterocles senegalus. In *Morokko. Journ. Für Ornith*. 110:181–191.

————. 1976. *In den Wüsten der Erde*. Hamburg.

George, W. 1987. Complex origins. In T. C. Whitmore, ed., *Biogeographical evolution of the Malay Archipelago*. Oxford.

Gerbert, M. 1970. *Religionen in Brasilien*. Bibliotheca Ibero-Americana, Vol. 13. Berlin.

Gerken, B. 1995. Die Weser von morgen, Part 1: Entwicklung eines Leitbildes für die Weser auf anthropologischer und naturgeschichtlicher Grundtage. In Gerken and Schirmer, eds., *Die Weser. Limnologie aktuell,* Vol. 6. Stuttgart.

Gerster, G. 1975. *Der Mensch auf seiner Erde—Ein Flugbild.* Zurich, Freiburg.

Gibbs, G. W. 1980. *New Zealand butterflies.* Auckland/Sydney/London.

Gieseler, W. 1974. Die Fossilgeschichte des Menschen. In G. Heberer, ed., *Die Evolution der Organismen,* Vol. 3, 3rd ed. Stuttgart.

Göbel, T. 1976. *Feuer-Erde.* Stuttgart.

Goethe, J. W. v., 1817. Zur Morphologie (zert nach D.Erstausgabe v. Goethes Naturwiss. Schriften durch R. Steiner, Vol. 1, Stuttgart/Berlin/Leipzig 0. J.).

Goffart, M. 1971. *Function and form in the sloth.* New York.

Gómez-Pompa, A., and A. Kaus. 1990. Traditional management of tropical forests in Mexico. In A. B. Anderson, ed., *Alternatives to deforestation—steps toward sustainable use of the Amazon rain forests,* 45–64. Columbia University, New York/Oxford.

Gómez-Pompa, A., C. Vasques-Yanes, and S. Guevara. 1973. Der tropische Regenwald: Von endgültiger Vernichting bedroht. *Umschau* 73, Nr. 16:300.

Goodland, R., and H. S. Irwin. 1975. *Amazon jungle: green hell to red desert?* Amsterdam.

Grohmann, G. 1931. *Entwicklungsgesetze in der fossilen Pflanzenwelt.* In *Gaia Sophia,* Jahrbuch der Naturwiss. Sektion der Freien Hochschule am Goetheanum. Dornach.

Grossbach, I., and W. Schad. 1974. Niedermoor und Hochmoor. *Elemente der Naturwissenschaft* Nr 21:22–40.

Gümbel, D. 1972. Biologische Seentypen und die Gesetzmäßigkeit ihrer Verbreitung (Entwurf zu einer Organologie der Erde). Unveröff Manuskript.

Gut, B. 1974. Zur Idee des Goetheanum-Geländes. *Die Drei* 44:370–381.

Harlan, J. R. 1977. The origins of cereal agriculture in the Old World. In C. Reeds, ed., *The Origins of Agriculture.* The Hague.

Heberer, G. 1974. Die subhumane Abstammungsgeschichte des Menschen. In G. Heberer, ed., *Die Evolution der Organismen,* Vol. 3, 3rd ed. Stuttgart.

Hedberg, O. 1951. Vegetation belts of the East African mountains. *Svensk bot. Tidskr.* 45:140–202.

————. 1961. The phytogeographical position of the afroalpine flora. *Recent advances in botany,* 914–919. Toronto.

Hediger, H. 1948. Kleine Tropenzoologie. *Acta Tropica* Suppl. 1. Basel.

————. 1950. La capture des éléphants au Parc National de la Garamba. Bull .Inst. Roy. Col. Belge, Vol. 21.

Heezen, B. C. and I. D. MacGregor. 1973. The evolution of the Pacific. *Scientific American*, Nov.

Heim de Balsac, H. 1936. Biogéographie des mammifères et des oiseaux de l'Afrique du Nord. *Bull. Biol.*, Suppl. 21.

Hendrichs, H. 1970. Schätzungen der Huftierbiomasse in der Dornbuschsavanne nördlich und östlich der Serengetisteppe in Ostafrika . . . Säugetierkundl. *Mitteilungen* 18:237–255.

————. and U. Hendrichs. 1971. *Dikdik und Elefanten. Ökologie und Soziologie zweir afrikanischer Huftiere.* Munich.

Herlocker, D. J., and H. J. Dirschl. 1972. *Vegetation of the Ngorongoro Conservation Area, Tansania.* Canad. Wildlife Service Report Series No. 19.

Hirschberg, W. 1965. *Völkerkunde Afrikas.* Mannheim.

Hochstetter, F. v. 1867. *New Zealand, its physical geography, geology and natural history with special reference to the results of government expeditions in the Provinces of Auckland and Nelson.* Stuttgart.

Hölldobler, B., and E. O. Wilson. 1990. *The ants.* Berlin, Heidelberg.

Howell, J. M. 1989. Jungsteinzeitliche Agrikulturen in Nordwesteuropa. In *Siedlungen der Steinzeit, Spektrum d. Wissenschaft, Reihe: Verständliche Forschung.* Heidelberg.

Hueck, K. 1966. *Die Wälder Südamerikas.* Stuttgart.

Huttel, C. 1975. Root distribution and biomass in three Ivory Coast rain forest plots. In F. B. Golley and E. Medina, eds., *Tropical ecological systems.* Berlin, Heidelberg, New York.

Huxley, J. S. 1954. The evolutionary process. In J. S. Huxley and A. C. Hardy, eds., *Evolution as a process.* London.

————. 1965. Die Reichtümer des wilden Afrika. In *Ich sehe den zukünftigen Menschen.* Munich.

Keay, R. W. J. 1959. Vegetation map of Africa south of the tropic of cancer. London.

Kemp., E. M. 1981. Tertiary palaeogeography and the evolution of Australian climate. In A. Keast, ed., *Ecological biogeography of Australia.* The Hague/ Boston/London.

Kingdon, J. 1971–1977. *East African mammals.* Vols 1–3.

Kipp, F. A. 1955. Die Entstehung der menschlichen Lautbildungsfähigkeit als Evolutionsproblem. *Experintia,* Vol. 11, 89–94. Stuttgart.

———. 1985. Indizien für die Sprachfähigkeit fossiler Menschen. In W. Schad, ed., *Goetheanistiche Naturwissenschaft,* Vol. 4, *Anthropologie.* Stuttgart.

Knapp, R. 1973. *Die Vegetation Afrikas.* Stuttgart.

Koehler, R. 1971. Der Flußlauf als ein Organismus. *Erziehungskunst* 35, Nr. 7/8/9.

Koepcke, H. W. 1973. *Die Lebensformen,* Vol. 1. Krefeld.

Kordtland, A. 1962. Chimpanzees in the wild. *Scientific American* 206:128–138.

Kumerloeve, H. 1975. *Die Säugetiere (Mammalia) der Türkei.* Veröff. Zool. Staatssamml. Munich.

Kurt, F. 1986. *Das Elefantenbuch. Wie Asiens letzte Riesen Leben.* Hamburg/Zurich.

Lachner, R. 1969. *Paradies der wilden Vögel Ostafrika.* Munich.

Landolt, E. 1970. Mitteleuropäische Wiesenpflanzen als hybridogene Abkömmlinge von mittel- und südeuropäischen Gebirgssippen und sub-mediterranen Sippen. *Feddes Repertorium* 81,I-5:61–66.

Lattin, G. de. 1967. *Grundriß der Zoogeographie.* Jena.

Lauer, W. 1989. *Ecosystems of the world,* 14 B. Amsterdam.

Lawick-Goodall, J. van. 1971. *Wilde Schimpansen. 10 Jahre Verhaltensforschung am Gombe-Strom.* Hamburg.

Laws, R..M. 1970. Elephants as agents of habitat and landscape change in East Africa. *Oikos* 21, 1–15.

Leakey, M. D. 1979. Footprints frozen in time. *National Geographic* 155:446–457.

———, and R. L. Hay. 1979. Pliocene footprints in the Laetoli beds at Laetoli, northern Tanzania. *Nature* 278:317–323.

———. and R. E. F. Leakey. 1978. Koobi Fora research project, Vol 1. *The fossil hominids and an introduction to their context 1968–1974.* Oxford.

Lenz, F. 1971. *Bildsprache der Märchen.* Stuttgart.

Leuthold, C. 1995. Die Ökogenese eines Landschaftsorganismus, dargestellt am Beispiel der Sukzession auf den Jungmoränen des Großen Aletschgletschers. Siehe in diesem Band S.

Lindroth, C. H., and G. H. Schwabe. 1970. Surtsey Island. Natürliche Erstbesiedlung (Ökogenese) der Vulkaninsel. Kiel.

Lockley, R. M., and N. W. Cusa. 1980. *New Zealand's endangered species*. Auckland.

Lovis, J. D. 1989. Timing, exotic terranes, angiosperm diversification and panbiogeography. *New Zealand Journ. Zool.* 16/4:713–729.

Lüning, J., and P. Stehli. 1989. Die Bandkeramik in Mitteleuropa: Von der Natur-zur Kulturlandschaft. In *Siedlungen der Steinzeit, Spektrum d. Wissenschaft, Reihe Verständliche Forschung*. Heidelberg.

Maes, J. 1951. Belgischer Kongo. In H. Bernatzik, ed., *Afrika, Handb. D. angew. Völkerkunde*, Vol. 2. Munich.

Mägdefrau, K. 1968. *Paläobiologie der Pflanzen*. 4th ed. Stuttgart.

Martin, R., and K. Saller. 1959. *Lehrbuch der Anthropologie*, Vol. 2. Stuttgart.

Maull, O. 1973. Zur Geographie Kulturlandschaft. In K. Paffen, ed., *Das Wesen der Landschaft*. Darmstadt.

McKay, G. M. 1973. *Behavior and ecology of the Asiatic elephant in southeastern Ceylon*. Smithson. Contrib. To Zoology 125. Washington.

McNeil, M. 1964. Lateritic soils. *Scientific American* 222, Nr. 11:97–102.

Mitchell, A. W. 1987. *The enchanted canopy—secrets from the rainforest roof*. Fontana/Collins, London.

Monod, T. 1958. Majâbat al-Koubra. Contribution à l'étude de l'"Empty Quarter" ouest-saharienn. Mem. Inst. Fr. Afr. Noire No. 52.

Montagu, A. 1984. *Growing young*. New York.

Moreau, R. E. 1952. Africa since the Mesozoic, with particular reference to certain biological problems. *Proc. Zool. Soc. London* 121:869–913.

———. 1963. Vicissitudes of the African biomes in the Late Pleistocene. *Proc. Zool. Soc. London* 142:395–421.

———. 1966. *The bird faunas of Africa and its islands*. London.

———. 1972. *The Palaearctic-African bird migration systems*. London.

Müller, S. 1984. Tiefenstruktur Dynamik und Entwicklung des Mittelmeer- und Alpenraunes. *Vieteljahresschrift Naturforsch. Ges. Zurich* 129:217–245.

Murdock, G. P. 1959. *Africa: its peoples and their culture history*. New York.

Nachtigall, G. 1879–1889. *Sahara und Sudan*, 3 vols. Leipzig.

Neumann, I. 1979. Beispiele ökologischer Landwirtschaft in den feuchten Tropen Mexikos. *Lebendige Erde* 4:138–144.

North, M. E, W. 1958. *Voices of African Birds.*

———, and D. McChesney. 1964 *More Voices of African Birds.* Recordings of the Laboratory of Ornithology, Cornell University. Ithaca, N.Y.

O'Neill, J. P., and D. L. Pearson. 1974. Estudio Preliminaire de las aves de Yarinacocha, Departmento de Loreto, Peru. Mus. Hist. Nat. "Javier Prado," Lima Zoological Serie A No. 25.

Owen, D. F. 1976. *Animal ecology in tropical Africa.* 2nd ed. London.

Papageorgis, C. 1975. Mimicry in neotropical butterflies—Why are there so many complexes in one place? *American Scientist* 63:522–532.

Pavoni, N. 1969. Zonen lateraler horizontaler Verschiebung in der Erdkruste und daraus ableitbare Aussagen zur globalen Tektonik. *Geol. Rundschau* 59:56–77.

Peppelbaum, H. 1931. *Man and animal.* Anthroposophical Publishing Co. London.

Poole, L., and N. Adams. 1990. *Trees and shrubs of New Zealand.* Wellington.

Portmann, A. 1948. *Einführung in die vergleichende Morphologie der Wirbeltiere.* Basel.

———. 1965. *Die Tiergestalt.* 2nd ed. Freiburg.

Prance, G. T., ed. 1986. *Tropical forests and the world atmosphere.* American Association for the Advancement of Science. Boulder, Colo.

Prospero, J. M., R. A. Glaccum, and R. T. Nees. 1981. Atmospheric transport of soil dust from Africa to South America. *Nature* 289:570–572.

Rahm, U. 1973. *Flora und Fauna des afrikanischen Regenwaldes.* Image Roche Nr. 49 and 53. Basel.

Recher, H. F., and P. Christensen. 1981. Fire and the evolution of the Australian biota. In A. Keast, ed., *Ecological biogeography of Australia.* The Hague/Boston/London.

Reichholf, J. H. 1990. *Der tropische Regenwald. Die Ökobiologie des artenreichsten Naturraumes der Erde.* Munich.

Remane, A. 1950. Ordnungsformen der lebenden Natur. *Studium Generale* 3:404–410.

Reynolds, V. 1966. *Budongo.* Wiesbaden.

Richard, J.-L. 1968. *Les groupements végéde la reserve d'Aletsch.* Bern.

Richards, P. W. 1964. *The Tropical Rain Forest.* Cambridge.

————. 1973. Africa, the "odd man out." In B. J. Meggers, E. S. Ayensu, and W. D. Duckworth, eds., *Tropical forest ecosystems in Africa and South America: a comparative review,* 21–26. Washington.

Ruhberg, H. 1985. Die Peripatopsidae (Onychophora). *Zoologica* 46/137.1

Samwald, J., and Y. Ditfurth. 1989. Kleiner Garten Eden. *Natur* 12:38–46.

Schad, W. 1974. Erziehungskunst aus anthroposophischer Menschenkunde. *Erziehungskunst* 38:233–237.

————. 1977. *Man and mammals: toward a biology of form.* Trans. Carroll Scherer. Waldorf Press, New York.

————. 1985. Gestaltmotive fossiler Menschenformen. In W. Schad, ed., Goetheanische Naturwissenschaft, Vol. 4, *Anthropologie,* 57–152. Stuttgart.

————. 1992. Der Heterochronie-Modus in der Evolution der Wirbeltierklassen und Hominiden. Dissert., unveröff. Manuscr. Univ. Witten-Herdecke.

————. 1996. Urgeschichtkiches Israel—Schwelle und Durchgangsland der Menscheitsentwicklung. In A. Suchantke, ed., *Mitte der Erde. Israel in Brennpunkt natur—und kulturgeschichtlicher Entwicklungen.* 2nd. ed. Stuttgart.

Schaller, G. 1963. *The mountain gorilla.* Chicago, London.

————. 1967. *The deer and the tiger.* Chicago, London.

————. 1972. *The Serengeti lion.* Chicago, London.

Schebesta, P. 1938. Die Bambuti-Pygmäen von Ituri. Inst. Roy. Colon. Belge Bruxelles, Sect. Science morales et politiques, Memoires, Coll. In 4∫, Vol. I.

Schenkel, R., and. L. Schenkel-Hulliger. 1969. *Ecology and behaviour of the black rhinoceros (Diceros bicornis).* Mammalia depicta, Hamburg and Berlin.

Schindewolf, O. 1950. *Grundfragen der Paläontologie.* Stuttgart.

————. 1972. Phylogenie und Anthropologie aus paläontologischer Sicht. In H. G. Gadamer and P. Vogler, eds., *Neue Anthropologie,* Vol. 1, Part 1. Stuttgart.

Schmithüsen, J. 1973. Was ist eine Landschaft? In K. Paffen, ed., *Das Wesen der Landschaft.* Darmstadt.

Schmutz, H. U. 1986. *Die Tetraederstruktur der Erde. Eine geologisch-geometrische Untersuchung anhand der Plattentektonik.* Stuttgart.

————. 1996. Zur Geologie Palästinas. In A. Suchantke, ed., *Mitte der Erde. Israel in Brennpunkt natur—und kulturgeschichtlicher Entwicklungen.* 2nd. ed. Stuttgart.

Schwidetzky, I. 1974. Rassenevolution des Menschen. In G. Heberer, ed., *Die Evolution der Organismen,* Vol. 3. 3rd ed. Stuttgart.

Sergejew, M. 1986. *Irkutsk.* Moscow.

Shortridge, G. C. 1934. *The mammals of South West Africa.* 2 vols. London.

Sibatani, A. 1974. A new species of *Lycaeninae (s. str.) (Lepidoptera Lycaenidae)* from Papua New Guinea. *Journ. Austral. Entomol. Soc.* 13:95–110.

Sibley, C. G., G. R. Williams, and J. E. Ahlquist. 1982. The relationships of the New Zealand wrens (Acanthisittidae) as indicated by DNA-DNA hybridisation. *Notornis* 29:113–130.

Sick, H. 1958. *Tukani. Unter Tieren und Indianern Zentralbrasiliens.* Hamburg, Munich.

Sioli, H. 1973. Recent human activities in the Brazilian Amazon region and their ecological effects. In B. J. Meggers, E. S. Ayensu, and W. D. Duckworth, eds., *Tropical forest ecosystems in Africa and South America: a comparative review,* 321–334. Washington.

Smith, J. M. B. 1986. Origin and history of the Malesian mountain flora. In F. Vuilleumier and M. Monasterio, eds., *High altitude tropical biogeography.* New York/Oxford.

Smith, M. H. 1988. *Wild south: saving New Zealand's endangered birds.* Auckland.

Steiner R. 1910. *Die Mission einhinzelner Volksseelen in Zusammenhange mit der germanisch-nordlischen Mythologie.* 4th ed. Dornach, 1962.

———. 1916. *The riddle of humanity.* London.

———. 1917. *Metamorphosen des Seelenlebens.* 4th ed. Dornach, 1958.

———. 1918. Soziale und antisoziale Triebe in Menschen. In *Die soziale Grundforderung unserer Zeit—In geänderter Zeitlage.* Dornach, 1963.

———. 1924. *Spiritual foundations for the renewal of agriculture,* Kimberton, Penn., 1993.

Steward, J. H. 1948. *Handbook of South American Indians,* vol. 3, *The tropical forest tribes.* Washington.

Stewart, W. N. 1983. *Paleobotany and the evolution of plants.* Cambridge Univ. Press.

Suchantke, A. 1965. *Metamorphosen im Insektenreich.* Stuttgart.

———. 1972. *Sonnensavannen und Nebelwälder.* Stuttgart.

———. 1974. Afrikanische Begegnungen. *Die Drei* 44:268–273.

———. 1974. Biotoptracht und Mimikry bei afrikanischen Tagfaltern. *Elemente der Naturwissenschaft* 21:1–21.

————. 1976. Biotoptracht bei sudamerikaniscchen Tagfaltern. *Elemente der Naturwissenschaft* 251–258.

————. 1982. *Der Kontinent der Kolibris. Landschaften und Lebensformen in den Tropen Sudamerikas.* Stuttgart.

————. 1986. Die Landschaft des Menschen. Eichen-Savannen und Wildgetreide-fluren in Palästina. *Die Drei* 56/4:259–278.

————. 1987. Mensch und Natur in anderen Kontinenten und Kultur. *Die Drei* 5:345–356.

————. 1990. Das Problemfeld Mensch-Natur: Gibt es Chancen zur Kooperation? In M. Faulstich and K. E. Lorber, eds., *Ganzheitlicher Umweltschutz.* Edition Universitas, Stuttgart.

————. 1993. *Partnerschaft mit der Natur—Entscheidung für das kommende Jahrtausend.* Stuttgart.

————. 1995. The metamorphosis of plants as an expression of juvenilisation in the process of evolution. In J. Bockemühl and A. Suchantke, *The metamorphosis of plants.* Novalis Press, Cape Town.

————. 1996. Natur in Israel—Brennpunkt und Synthese weltweiter Einflüsse. In A. Suchantke, ed., *Mitte der Erde. Israel in Brennpunkt natur—und kulturgeschichtlicher Entwicklungen.* 2nd. ed. Stuttgart.

Sutherland, C. 1990. *Die Prinzessin von Sibirien—Maria Wolkonskaja und ihre Zeit.* Frankfurt a. M.

Takhtajan, A. 1991. *Evolutionary trends in flowering plants.* New York.

Talbot, L. M. 1963. The high biomass of wild ungulates on East African savanna. Trans. N. Amer. Wildlife Conf. 28. Toronto.

Taylor, C. R. 1969. The eland and the oryx. *Scientific American* 220:88–91.

Thienemann, A. F. 1956. *Leben und Umwelt. Vom Gesamthaushalt der Natur.* Hamburg.

Thomas, B. A., and R. A. Spicer. 1987. *The evolution and palaeobiology of land plants.* London.

Thorne, R. F. 1973. Floristic relationships between tropical Africa and tropical America. In B. J. Meggers, E. S. Ayensu, and W. D. Duckworth, eds., *Tropical forest ecosystems in Africa and South America: a comparative review,* 27–47. Washington.

Thorpe, W. H. 1973. Duet-singing birds. *Scientific American* 224:70–79.

Tischler, W. 1955. *Synökologie der Landtiere.* Stuttgart.

Troll, C. 1950. Die geographische Landschaft und ihre Erforschung. *Studium Generale* 3:163–181.

————. 1959. *Die Tropischen Gebirge.* Bonner Geogr. Abhand. 25.

————. 1969. *Vergleichende Morphologie der höherer Pflanzen.* Berlin 1937–1943—Die Inflorenszenzen. Stuttgart, 1964, 1969.

————. 1973. Landschaftsökologie als geographisch-synoptische Naturbetrachtung. In K. Paffen, ed., *Das Wesen der Landschaft.* Darmstadt.

Uerpmann, H. P. 1983. Die Anfänge von Tierhaltung und Pflanzenanbau. In H. Müller-Beck, ed., *Urgeschichte in Baden-Würtemberg.* Stuttgart.

van der Post, Laurens. 1955. *The dark eye in Africa.* William Morrow, New York.

van Steenis, C. G. G. J. 1964. Plant geography of the mountain flora of Mt. Kinabalu. *Proc. Royal Soc. London* Ser. B 161:7–38.

Vareschi, V. 1980. *Vegetationsökologie der Tropen.* Stuttgart.

Vaucher, C. 1972. *Nakuru, See der Flamingos.* Zurich, Morges.

Veltheim-Ostrau, H. H. v. 1954. *Der Atem Indiens.* Tagebücher aus Asien, *Neue Folge: Ceylon und Südindien.* Hamburg.

Vincent, J. 1974. The management of large mammals in Natal, with special reference to utilization for stocking and restocking. In V. Geist and F. Walther, eds., *The behaviour of ungulates and its relation to management.* IUCN Publications new series No. 24:900–908.

Vogel, S. 1972. Komplementarität der Biologie und ihr anthropologischer Hintergrund. In H. G. Gadamer and P. Vogel, eds., Neu Anthropologie, Vol. 1, *Biologische Anthropologie,* Part 1, 152–194. Stuttgart.

Vos, A. de. 1975. *Africa, the devastated continent?* The Hague.

Walker, A., and R. E. F. Leakey. 1978. The hominids of East Turkana. *Scientific American* 239, No. 2, 44–56.

Wallace, A. R. 1869. *The Malay Archipelago.* London.

Walter, H. 1939. Grasland, Savanne und Busch der ariden Teile Afrikas in ihrer ökologischen Bedingtheit. *Jahrb.wiss.Botanik* 87:750–860.

————. 1964, 1973. *Die Vegetation der Erde in öko-physiologischer Betrachtung,* Vol. 1, *Die tropischen und sudtropischen Zonen.* 3rd ed. Stuttgart (1973).

————. 1968. *Die Vegetation der Erde in öko-physiologischer Betrachtung,* Vol 2, *Die gemäßigten und arktischen Zonen.* Stuttgart.

————. 1976. *Die ökologischen Systeme der Kontinente (Biogeosphäre).* Stuttgart.

————, and S.-W. Breckle. 1983. *Ökologie der Erde*, Vol. 1.

Wardle, P. 1991. *Vegetation of New Zealand.* Cambridge/New York/Sydney.

Webb, C. J., W. R. Sykes, and P. J. Garnock-Jones. 1988. *Naturalized pterido-phytes, gymnosperms, dicotyledons. Flora of New Zealand.* Vol. 4. Christchurch.

Weber, B. 1992. Der Baikal—Geographische und biologische Aspekte eines außergewöhnlichen Süßwassersees. *Natur und Museum* 122:101–125.

Williams, J. G. 1971. Artisten der Vögel von Nakuru- und Naivashasee. In *Säugetierre und seltene Vögel in den Nationalparks Ostafrikas.* Hamburg, Berlin.

Wilson, H. D. 1976. *Vegetation of Mount Cook National Park, New Zealand.* National Parks Authority, Scientific Series 1. Wellington.

WWF News. 1987. The Kayapó Indians of Brazil: effective forest managers. WWF News 49, Sept./Oct. *WWF International,* Gland, Switzerland.

Zimmermann, W. 1952. Main results of the "telome theory." *Palaeobotanis* I:456–470.

————. 1965. *Die Telomtheorie.* Stuttgart.

————. 1969. *Geschichte der Pflanzen.* 2nd, rev. ed. Stuttgart.

Zohary, D. 1969. The progenitors of wheat and barley in relation to domestication and agricultural dispersal in the Old World. In P. J. Ucko and G. W. Dimbleby, eds., *The domestication and exploitation of plants and animals.* London.

————, and M. Hopf. 1993. *Domestication of plants in the Old World.* Oxford.